Product Design

Edited by Cătălin Alexandru, Codruta Jaliu and Mihai Comşit

Published in London, United Kingdom

IntechOpen

Supporting open minds since 2005

Product Design
http://dx.doi.org/10.5772/intechopen.83170
Edited by Cătălin Alexandru, Codruta Jaliu and Mihai Comşit

Contributors
Michael Lasithiotakis, Abas Haidari, Panagiotis Sinioros, Ourania Tzoraki, Nikolaos Manousakis, Rosnani Ginting, Hrayr Darbinyan, Catalin Alexandru, Wahiba Yaïci, Evgueniy Entchev, Pouyan Talebizadeh Sardari, Michela Longo, Ana Pilar Valerga Puerta, Severo R. Fernandez-Vidal, Sergio Fernandez-Vidal, Rocio A. Jimenez-Rodriguez, Aghaegbuna Obinna U. Ozumba, Winston Shakantu

First published in London, United Kingdom, 2020 by IntechOpen
IntechOpen is the global imprint of INTECHOPEN LIMITED, registered in England and Wales, registration number: 11086078, 5 Princes Gate Court, London, SW7 2QJ, United Kingdom
Printed in Croatia

British Library Cataloguing-in-Publication Data
A catalogue record for this book is available from the British Library

Additional hard and PDF copies can be obtained from orders@intechopen.com

Product Design
Edited by Cătălin Alexandru, Codruta Jaliu and Mihai Comşit
p. cm.
Print ISBN 978-1-83968-212-4
Online ISBN 978-1-83968-213-1
eBook (PDF) ISBN 978-1-83968-214-8

For EU product safety concerns:
IN TECH d.o.o., Prolaz Marije Krucifikse Kozulić 3, 51000 Rijeka, Croatia, info@intechopen.com or visit our website at intechopen.com.

Meet the editors

Cătălin Alexandru is a professor at the Transilvania University of Braşov (Romania), within the Faculty of Product Design and Environment. His teaching and research activity centers mainly on the modeling, simulation and optimization of mechanical and mechatronic systems. Prof. Alexandru is a PhD supervisor in Mechanical Engineering, and he is currently the director of the University Doctoral Council. He is also the coordinator of the Virtual Prototyping Laboratory at the Design of Mechanical Elements and Systems Research Center. He has published more than 300 scientific papers in mainstream journals, books chapters, and conference proceedings. He served as reviewer for dozens of ISI/WOS-rated journals. Prof. Alexandru is a member in national and international organizations that promote research in the field of product design.

Dr. Codruta Jaliu is professor in the Product Design and Environment Faculty of the Transilvania University of Brasov, as well as a PhD supervisor in Mechanical Engineering. Her research interests include mechanical and mechatronic systems design in general, and design of wind turbines and small hydropower plants. She is author of five patents and more than 170 papers published in mainstream journals, international publishing houses, and international conference proceedings. She is editor of the *Bulletin of the Transilvania University of Brasov, Series I. Engineering Sciences*, a member of the ARoTMM, the Roumanian branch of the International Federation for the Promotion of Mechanism and Machine Science (IFToMM), and reviewer for several journals. She is the coordinator of the Industrial Design bachelor program, guiding the students throughout each phase of the program, providing them administrative support and updating the curriculum according to the employers' needs.

Mihai Comşit is Associate Professor at the Transilvania University of Braşov (Romania), Faculty of Product Design and Environment. His teaching and research activity focus mainly on product design combining design, modeling, simulation, and optimization of mechanical systems expertise with artistic skills like drawing, painting, computer graphics, and computer-aided design, subjects that he also teaches at the university. His research interests include mechanical and mechatronic systems design in general, design of architecturally integrated solar energy conversion systems, and electric mobility. He is author of three patents and more than 50 papers in mainstream journals, international publishing houses, and international conference proceedings. He has also co-authored three books. He is the coordinator of the BlueStreamline Formula Student Racing Team of Transilvania University of Brasov achieving international prizes in this activity. He is also a founding member member of the ARoTMM, the Romanian branch of the International Federation for the Promotion of Mechanism and Machine Science (IFToMM), and a reviewer for several journals.

Contents

Preface

Product design is a comprehensive process related to the creation of new products. It starts from a set of ideas and concepts, and finally results in marketable products that meet specific requirements related to functionality, manufacturing, maintenance, reliability, ergonomics, efficiency, and, last but not least, cost. The design and manufacture of a product, whether new or existing, is a permanent concern and challenge. The ability to design and develop efficient products are key to success in today's dynamic global market.

This book is an open platform designed to establish and share knowledge developed by scholars, scientists, and engineers from all over the word about the product design process and its applications in various fields, especially those in engineering. The book is a forum for original and innovative research studies dealing with any relevant aspect of product design. It is especially dedicated to the most recent techniques for successful product design as well as to the most suitable software applications.

The book contains seven original chapters that provide a comprehensive overview of the current state of the art and modern trends in the field, reflecting the multidimensionality of the applications related to product design. Topics covered include development of new product design methodologies, implementation of effective methods for integrated products, development of more visualized environments for task-based conceptual design methods, development of engineering design tools based on 3D photogrammetry, adaptation of Information and Communication Technology (ICT) tools for improving the construction site management process, and the use of virtual prototyping software platforms.

The book was edited by a group of scientists from the Department of Product Design, Mechatronics and Environment, Faculty of Product Design and Environment, Transilvania University of Brașov (Romania), namely, Professors Cătălin Alexandru and Codruța Jaliu and Associate Professor Mihai Comsit.

Prof. Dr. Eng. Cătălin Alexandru, Codruta Jaliu and Mihai Comşit
Department of Product Design, Mechatronics and Environment,
Faculty of Product Design and Environment,
Transilvania University of Brașov,
Brașov, Romania

Section 1

Product Design Methods

Chapter 1

Task-Based Conceptual Design of a Novel Product

Hrayr Darbinyan

Abstract

A novel task-based conceptual design method introduced around a decade ago has been presented from its most characteristic points including the general idea, usage and modification of previous art, usage and modification of independent sets of functional and mechanical means for implementation of those functions, creation of intermediate mechanical-functional sets supporting the development of new structures like models, visualization of the design process, and so on. The current paper aims to reveal a non-computerized graphically visualized set of actions covering all the above-mentioned major steps of the suggested methodology. The success of synthesizing action greatly depends on the method of creation submechanisms or virtual mechanisms, which are making possible visualization and consideration intermediate structures helping to identify and implement a necessary function. The method of creating of such subcategories and application of elementary movements or set of links for explaining or satisfying demanded set of functions could be considered the main methodical novelty and strength of proposed conceptual design method. Two examples are included: the first reinvention of a known tool—Locking Pliers from database and second synthesis of a novel hand tool—Adjustable Nut Wrench.

Keywords: conceptual, design, task, structure, modification

1. Introduction

The conceptual phase of design still remains as the most challenging and less understood steps of general mechanical design. The difficulties of its description and formulization are coming from first the nature of conceptual design, implying search of a novel structure with novel properties among theoretically large number of candidate solutions, and secondly from the individual nature of designing process, depending on design tradition, skill, and experience of the designer. Design process being very attractive and creative by its definition may bring the designer more results and satisfaction if organized in a way to free the designer from the routine task of checking a large number of options with necessary efforts of visualization. The designer will greatly win if only decision-making duty from a limited number of candidate solutions will be left on his side. Combinational methods widely used for novel structure solution search are effective for the automatic organization of search process with minimum human involvement; however they are providing solutions for a key or major function, only suggesting modifications within fixed topology with randomly generated accompanying functions with a high probability of negative functions among them. The fact of single or key function consideration in combinational search dramatically lowers its methodical strength because any design process is valued for providing a multifunctional solution but not valued for the fact of generation from a single topology. A novel conceptual design is generally preceded by an act of making a decision which is possible after managing large data of previous knowledge and search of candidate solutions of the current design process. The success of large data management depends on the success of turning large data into small-sized easily manageable portions of information or models. From category point, the models should serve both categories involved, namely, functions and mechanism to provide interdependence and simultaneous consideration of those two, and from application point, they need to serve both classic actions of design—synthesis and analysis—including such segments of design as database analysis and creation of supplementary virtual mechanisms with the latest further update into structures, satisfying the given design tasks. That's practically an impossible task to understand and manage a designer's own plan during his or her efforts to create a novel mechanism. Commonly a designer who puts an aim to create a new mechanical structure or to update an existing one relies on proved by own practice and experience approaches and scenarios which could be basically different from each other and normally not shared with designers' community. For the past few decades, due to growing demand on fresh products with advantageous proprieties and because of wider application of digital technologies, the challenge of better organization of conceptual design process becomes more actual, and this demand was satisfied by several approaches and methodologies. The task-based design methods can be conventionally divided into methodologies based mostly on human participation or on computer-aided methods with minimum involvement of human factor. Some examples for the second group of task-based design methodologies are quite successful when directing a designer to organize a new product development with novel properties [1–3]. Very popular and classical methods [4, 5] of splitting mechanical components from functional ones have clear abstraction and visualization means and require consideration of a large number of candidate solutions in an attempt to isolate a workable and optimal one. A fundamental publication [6] is using analyses of the vast engineering database as a source for a novel product design, where the search trend implies consideration of either combination of various movements of basic links or direct search of solutions among existing solutions. Insufficient level of abstraction and visualization narrows the opportunities of processing and getting optimal results among mechanical means, having required functions and properties. Any design methodology can be evaluated by the number of essential design tasks considered during a mechanism synthesis process and distribution of those tasks along with steps of conceptual design: more tasks involved provides wider and full satisfaction of design aim with a maximum number of demanded properties of a novel product. When following [4, 5] methodologies, there is a great risk of missing and/or canceling consideration of essential features on one side and necessity of implementing of an exhausting search of candidate solution on the other. The largely popularized method of Theory of Inventive Problem Solving (TRIZ/TIPS) is quite effective for finding solutions of conceptual design and resolving invention tasks [7, 8]. A contradiction matrix and set of 40 creative/inventive tools are used for developing special auxiliary structures—VEPOLS (a Russian abbreviation of substance + field)—as a model of future novel product for revealing new and neutralizing the harmful functions. That's very common in mechanical engineering practice that an innovative solution surfs out once a limited set of various requirements are considered simultaneously, thus facilitating the creation of a decision-making situation. In the TIPS case, the fact of usage of a cumbersome set of 40 tools and no necessary relation of this set to the essence of the main problem generally may lead to comprehensive search with a broad number of possible results.

A task-based methodology of conceptual design [9] developed as a result of long-term engineering experience has proven its efficiency in the development of numerous and various mechanical devices based on interdependent and direct consideration of two sets of components—mechanical and functional—in a state when those components are processed to design models, facilitating their application or further modification for satisfying a current design task, while the remaining ones are planned to be satisfied by similar modification actions. Work models are developed at consecutive steps of the design process which may have different contents depending on the step, level, and scale of the design task.

In Sections 2.1 and 2.2, flat graphs are used for describing key mechanical and functional models; then the same graphs are used for visualizing modification (expansion and squeezing) of those models for serving different design needs. In Section 2.3 the development of local mechanical and functional models is presented; Section 2.4 covers developments for a solution means, namely, resources of database and resources of synthesis tools and similar developments for functional sets. Section 3 is, for example, relating to the reinvention of a known tool—locking pliers—from patent database, and finally, Section 4 is for the set of design cycles of conceptual design for a specific hand tool: self-adjustable nut wrench.

1.1 Tasks and objectives

The task of this paper is to provide clear graphical presentation and visualization for steps of the proposed approach of conceptual design methodology through mechanisms, functional-mechanical graphs, and set of hierarchized functions to control and manage the process of conceptual design, starting from the task on it and finishing with novel structure satisfying the pre-given tasks on development.

2. Main ideas of conceptual design method described by mechanical-functional graphs

2.1 Key models containing mechanical and functional categories

For mechanical categories the graph (**Figure 1**) has two vertices for links *L1* and *L2* related to each other by an edge as a kinematical joint in general or as another relation *R12* in a way that this relation may satisfy or plan satisfaction of demanded function *F12*. In fact, the links *L1* and *L2* are connected through two paths, firstly the edge *F12* represents the function as a subject of satisfaction, while the edge *R12* represents the mechanical or physical means needed for satisfying the demanded function. At the implementation stage, *F12* could be replaced by a physical kinematical joint shown in square symbol as *R12*.

For functional categories the graph (**Figure 2**) visualizes the step of getting a translated function *F2* from function *F1*, where edge *T12* stands for a translating operator, while the physical implementation of function *F2* is supposed to be done through mechanical means *M12*, represented by the second edge in the graph (**Figure 2**). That's easy to notice the topological analogy between **Figure 1** and **Figure 2**. Vertices of graph in **Figure 2** also are connected through two paths, firstly by the edge representing the translating operator *T12* providing a child function *F2* which may be implemented in contrary to function *F1* and secondly including second edge *M12* representing the mechanical mean for such implementation. *R12* in square symbol in **Figure 2** stands for the type of relation between functions *F1* and *F2*.

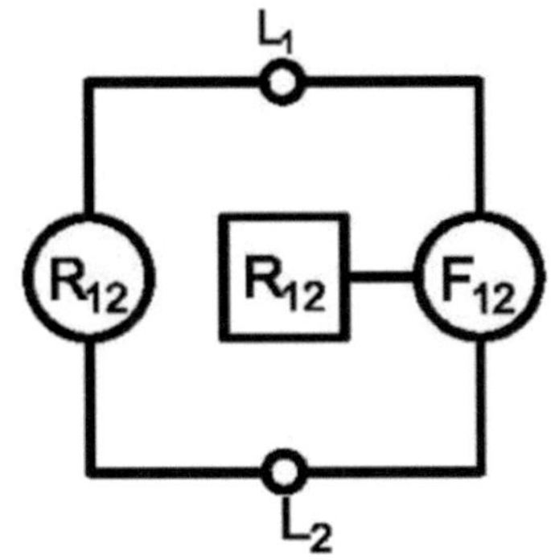

Figure 1.
Graph model for mechanical set.

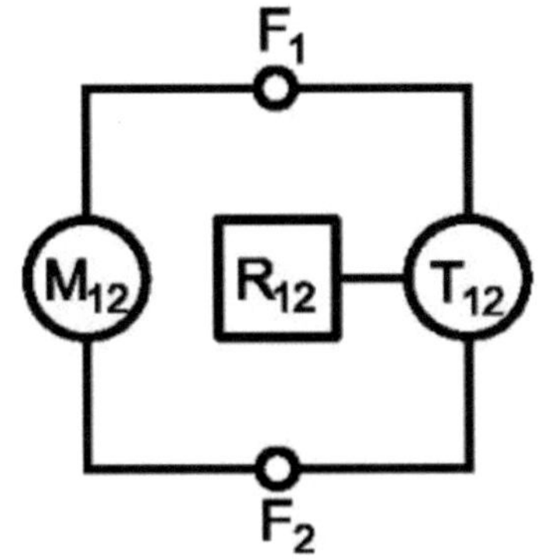

Figure 2.
Graph model for functional set.

It is worthy to note that both models include two initiating components, which in the mechanical model can be interpreted as a necessity of two links for getting a movement between them to implement a function. Analogously in the functional model, the presence of two functions can be interpreted as a necessity for having at least a function next to the initial function to provide its implementation by mechanical means. Generally, a single function also may serve as a task for development; anyhow a two-function model is considered to keep topological similarity of two basic models and also for presenting the translator operator which has the analogy to the relation of the links in the mechanical model in **Figure 3**.

The graphs in **Figures 1** and **2** are confirming the main idea of the proposed method of direct interdependence between mechanical and functional means for the fact of the presence of a functional edge in mechanical graph and presence of a mechanical edge in the functional graph. By the progress of the design process, the edges for functional graphs should be gradually substituted by physical relations between the links, and thus a novel mechanical structure should be generated at the end of design.

Structurally both design components' links and functions are centrally located in both graphs, leaving the left cells for relations between components and the right cells for the purpose of the means of implementation of those relations.

Other characteristic properties of the concept design method—interdependence of mechanical and functional categories and therefore possibility of combined consideration as a condition of target-oriented organization of conceptual design process—can be topologically visualized by the insertion of functional contents from

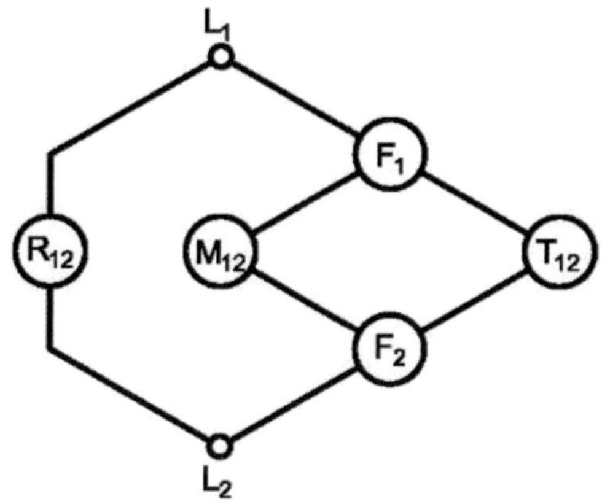

Figure 3.
Combined mechanical graph.

Figure 2 into the functional edge in **Figure 1**, thus getting combined mechanical functional graph (**Figure 3**), and correspondingly by inserting mechanical contents of the graph (**Figure 1**) into the mechanical edge of the functional graph (**Figure 2**), thus getting combined functional-mechanical graph (**Figure 4**).

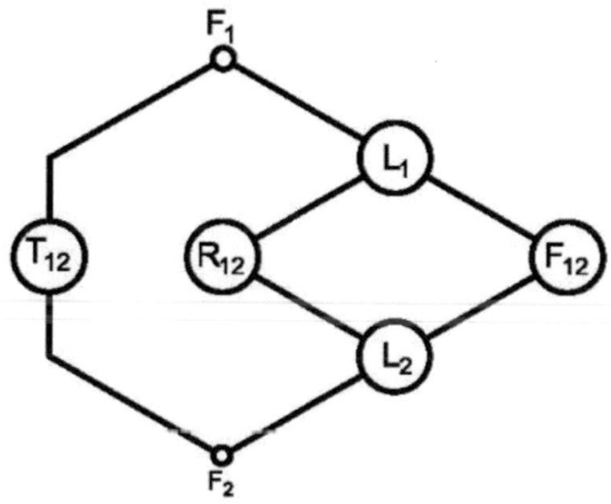

Figure 4.
Combined functional graph.

2.2 Models describing modification of mechanical and functional categories by means of expansion and squeezing

Once model graphs (**Figures 1–4**) are setting contents (centrally located) and relation (located on the left and right sides) of components of mechanical and functional matrices, expansion and squeezing actions aim to set synthesis preparation step. Those actions aim to establish the necessary scope of the search in a manageable way, avoiding exponential growth of components which leads to an exhausting search of sought solution. For the case of mechanical graph (**Figure 6**), the action of expansion results in necessary multiplication or addition of number of links in the central graph with an indication of mutual relation presented by a generalized symbol of *R18* (**Figure 5**). For the case of a functional graph, the same similar action leads to necessary multiplication or addition of the number of functions generated in an attempt to have the chance of their satisfaction by mechanical graph (**Figure 6**). Symbol *T18* stands for generalized translation operator between

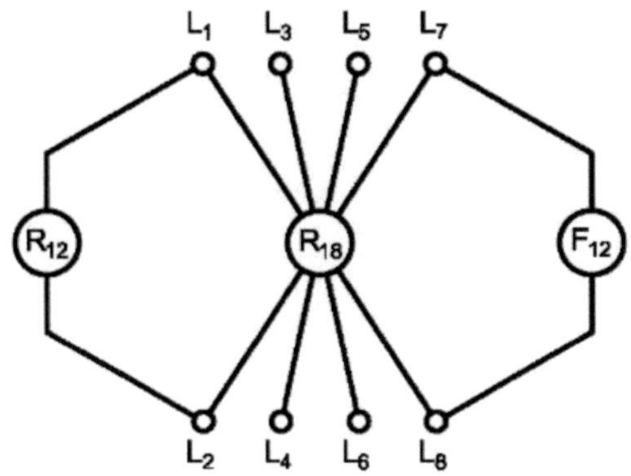

Figure 5.
Graph model for expansion and squeezing of mechanical means.

functions *F1 and F8*. Graphs (**Figures** 5 and **6**) can show the expandable and squeezable nature of both mechanical and functional entities by a simple modification—multiplication of number of links and multiplication of the number of functions. Along with multiplication, the edges of graphs should provide connections for both links and functions, thus confirming the function-based modification for the mechanical graph case and confirming the generation of new functions in the functional graph case. The opposite action of subtracting the links and functions will relate to squeezing case.

The combination of edges *F12* from **Figure** 5 and M12 from **Figure 6** helps to compose combined expanding-squeezing model as shown in **Figure** 7.

Represented modifications of both mechanical and functional means have the ability to disclose the hidden functional resources of the mechanical side and hidden ways of mechanical implementation of the functional side. Accordingly, both models have the ability of concentration of a limited number of links for mechanical side, creating local mechanical models and for a concentration of a limited number of functions for creating a local set of functions subject to implementation.

The case when the mechanical set consists of just one link may be interpreted as the squeezed down the state of set of links which were initially connected by a set of relations and then unified into a single link after consideration and implementation of those functional requirements. The mechanical set (**Figure** 5) can include the entire set of concept design components when the generalized function is substituted by its contents from **Figure 6**.

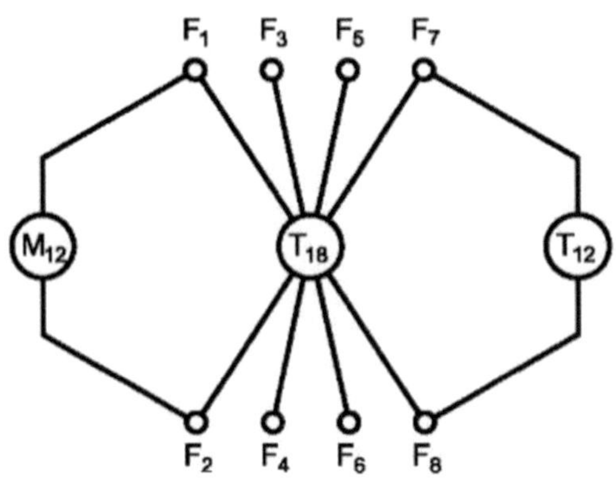

Figure 6.
Graph model for expansion and squeezing of functions.

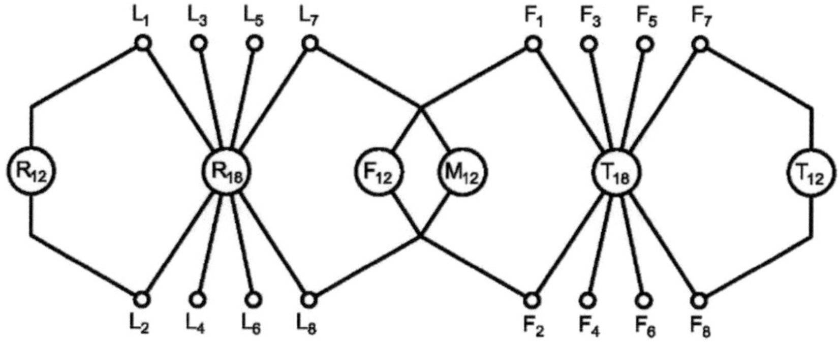

Figure 7.
Graph for combined expanding-squeezing mechanical-functional model.

2.3 Development of mechanical and functional models as preparation for synthesizing action

The modification of mechanical and functional sets is following different design goals: search and disclosure of hidden resources—in case of expansion and localization of the task and in case of squeezing. In both cases, interdependent portions of mechanical means and functions are being isolated. Those portions are easily manageable for analyzing the function implementation scenarios and for building a new structure with new properties. As described above isolated portions of mechanical/ functional sets are called models which are concentrating the main problem at a current design state, separating the problem from the general design process.

Localization and isolation of a problem are a widely practiced [3, 6, 7] action in mechanical design, supporting the task concentration, task targeting, and finding a solution. Anyhow the approach based on freedom and possibility of management and modification of two sets of mechanical and functional means in a direct and interdependent way, which opens a large-scale opportunity for finding an optimal solution; avoiding a large-scale exhausting search makes the proposed approach advantageously and effectively different from the abovementioned task localization approaches and design procedures. Thereby, two fragments of links *L3* ... *L6* and functions *F3* ... *F6* from **Figure 8** are separated and localized from general mechanical *(M)* and functional *(F)* sets. The isolated fragments thus are containing a low number of links and functions, and they are isolated in a way to focus designer attention on a finding of a solution of an isolated or targeted objective function. The isolated fragments of mechanical means contain links which either have the ability

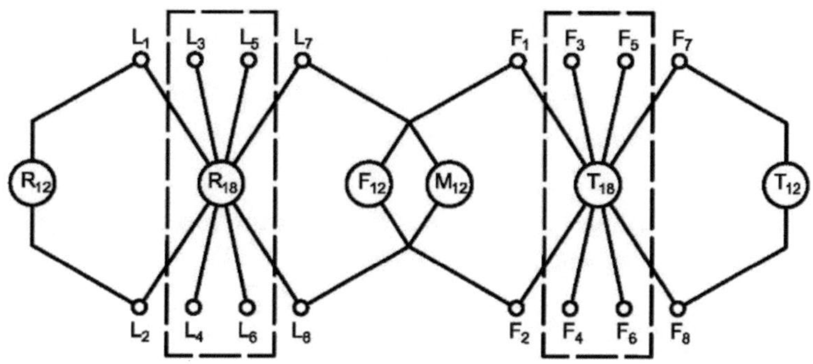

Figure 8.
Graph representation of two localized fragments (models) in mechanical and functional sets.

to satisfy the sought function or may gain this ability after proper modification, adding links, chains, grouping links, etc.

Localization of *mechanical and functional* fragments relates not only to neighbor components as it stated for convenience and simplicity but to any combination of components from **Figures** 5 and **6**.

2.4 Database modification and usage as a means for synthesizing action along with other synthesizing tools

The model presented in **Figure 8** serves the formulation of the problem and setting objective of synthesis which should be searched or prepared, depending on its availability in database or by satisfying challenging function by an initiated movement of a link. In other words, the first action relates to the search and to the usage of a confirmed function from database, which in wider interpretation can be considered as known mechanisms or fragment of mechanisms, and for the second case, setting a movable link originated from the basic one in a way to provide one of the common mechanical primitive functions (cam, screw, lever, gear, etc. mechanisms). The second action relates to a donation of degree of freedom action, where the movement of the novel originated link relative to the basic link may be interpreted in a wider range of functions than normal, in which it is interpreted by the designer. It is worthy to emphasize that same modeling techniques for mechanism and function (**Figures 1** and **2**), modification (**Figures 5** and **6**), and modeling (**Figure 8**) combined as a task block of conceptual design are identically applicable for a solution means. Corresponding models (**Figures 1, 2, 5, 6** and **8**) are valid for solution block, and those models could be developed using the same above-described procedure and modification techniques. Once a solution block is developed upon requirements of conceptual design, then this block could be aggregated with links of the preparation and task setting block. Graphically this aggregation is shown in **Figure 9** where two identical per-content and per-development procedure blocks stand on the left and right sides relative to centrally located aggregation symbols *AMD* (A for aggregation, M for mechanical block, D for database or solution block) which are representing edges of the graph, connecting corresponding links from task preparation block and solution block.

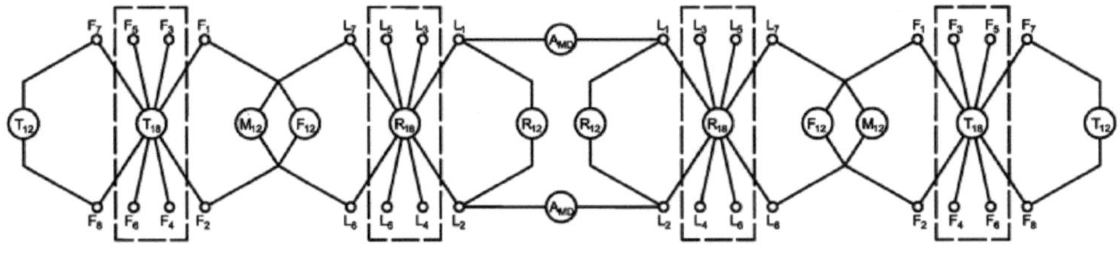

Figure 9.
Model for aggregation graph connecting task solution preparation block on the left and database or solution block on the right.

3. Application task-based conceptual design method on analyses and reinventing of a locking pliers (US patent, 1970)

3.1 Eight sets of design cycles

Eight sets of design cycles below track the concept design or redesign of a known tool locking pliers patented in the USA in 1970. The process starts with setting of objectives for design and ends with the structural diagram. Each design cycle is provided by explanations cited in **Figures 10–17**. Each design cycle shows the degree of implementation of design goal starting from 0 and ending with 1 (100%).

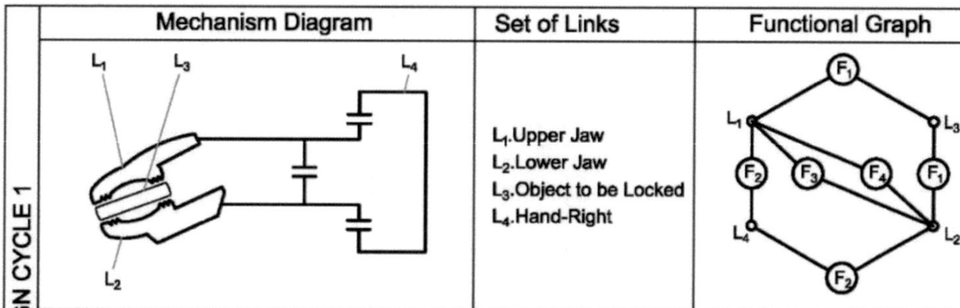

State of Functions

DESIGN CYCLE	Weight	1	2	3	4	5	6	7	8
F_1.Lock an Object	0.1	0							
F_2Hold in Hand	0.1	0							
F_3.Adjustable Thickness of Object	0.2	0							
F_4.Release	0.2	0							
F_5.Develop pressure	0.4	0							
Plan / Implementation	1 / 0	1 / 0							

Figure 10.
This is a task for synthesis; four basic functions are set as subject for implementation. Lock an object, provide adjustable thickness of locked object, develop necessary pressure for locking, and release the object if necessary. Functions are weighed by proportional values. Each design cycle shows status of a specific function implementation. Zero means function is planned but not implemented, empty cell means function is not planned, and the presence of a numerical value shows implementation.

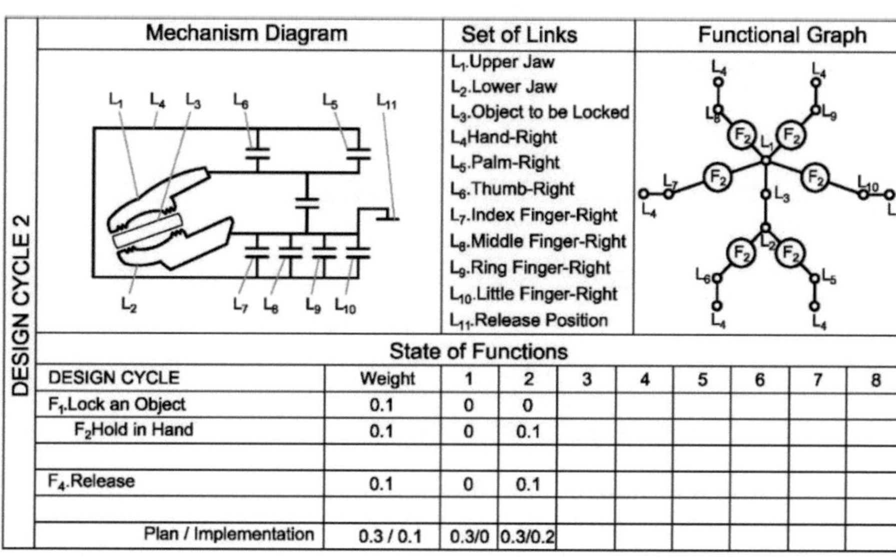

State of Functions

DESIGN CYCLE	Weight	1	2	3	4	5	6	7	8
F_1.Lock an Object	0.1	0	0						
F_2Hold in Hand	0.1	0	0.1						
F_4.Release	0.1	0	0.1						
Plan / Implementation	0.3 / 0.1	0.3/0	0.3/0.2						

Figure 11.
This design cycle is for planning such feature on convenient usage as holding in one hand while locking an object and releasing the object upon necessity.

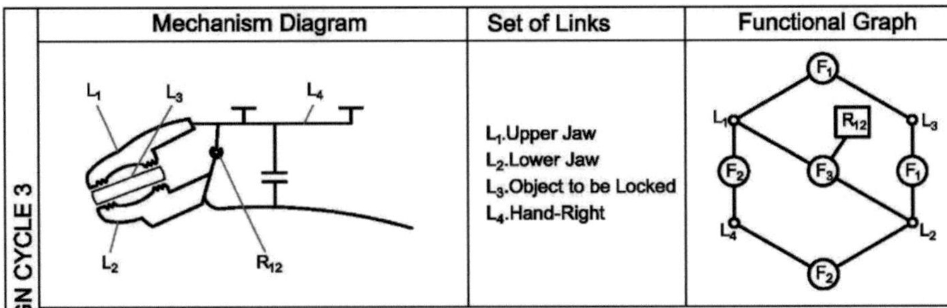

State of Functions

DESIGN CYCLE	Weight	1	2	3	4	5	6	7	8
F_1.Lock an Object	0.1	0	0	0					
F_2Hold in Hand	0.1	0	0.1	0.1					
F_3.Adjustable Thickness of Object	0.1	0		0.1					
F_4.Release	0.1	0	0.1	0.1					
Plan / Implementation	0.4/0.3	0.4/0	0.4/0.2	0.4/0.3					

Figure 12.
Adjustable feature as well the object locking feature are provided by pivoting Jaw1 against Jaw2, an action which can be explained also by an action of granting DOF.

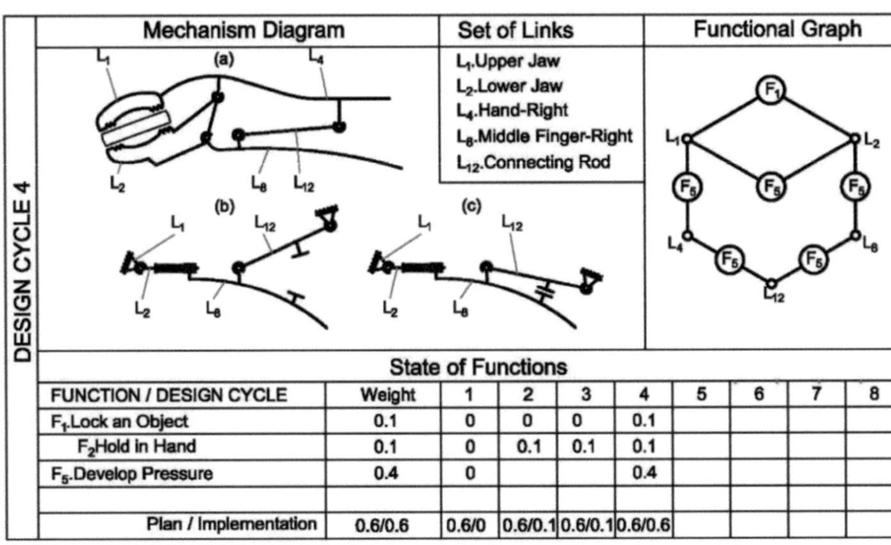

State of Functions

FUNCTION / DESIGN CYCLE	Weight	1	2	3	4	5	6	7	8
F_1.Lock an Object	0.1	0	0	0	0.1				
F_2Hold in Hand	0.1	0	0.1	0.1	0.1				
F_5.Develop Pressure	0.4	0			0.4				
Plan / Implementation	0.6/0.6	0.6/0	0.6/0.1	0.6/0.1	0.6/0.6				

Figure 13.
Property of developing pressure is implemented by an action of inserting a fragment of four bar mechanism, which at the region of end point is able to generate high force by lower amount of movement; a special virtual mechanism is developed (b, c) for explanation and confirmation of high transmission ratio feature.

4. Conceptual design example for a self-adjustable nut wrench

4.1 Task of conceptual design

The presented example of conceptual design refers to the search and finding a solution for a unique structure hand tool—a self and fast action adjustable socket—nut wrench [10]. According to its features, it replaces a large socket set for both metric and English sizes, differs in its features and outer look from any similar

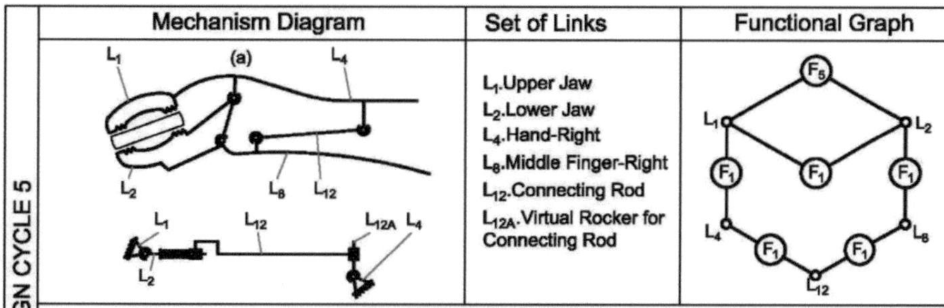

State of Functions

DESIGN CYCLE	Weight	1	2	3	4	5	6	7	8
F_1.Lock an Object	0.1	0	0	0	0.1	0.1			
F_2Hold in Hand	0.1	0	0.1	0.1	0.1	0.1			
F_5.Develop Pressure	0.4	0			0.4	0.4			
Plan / Implementation	0.6/0	0.6/0	0.6/0.1	0.6/0.1	0.6/0.6	0.6/0.6			

Figure 14.
Development of virtual mechanism for explaining the locking-fixing function of locking pliers. The virtual mechanisms provides confirmation and positive usage of locking force and neutralization of a force that attempts to release and cancel locked condition of locking pliers.

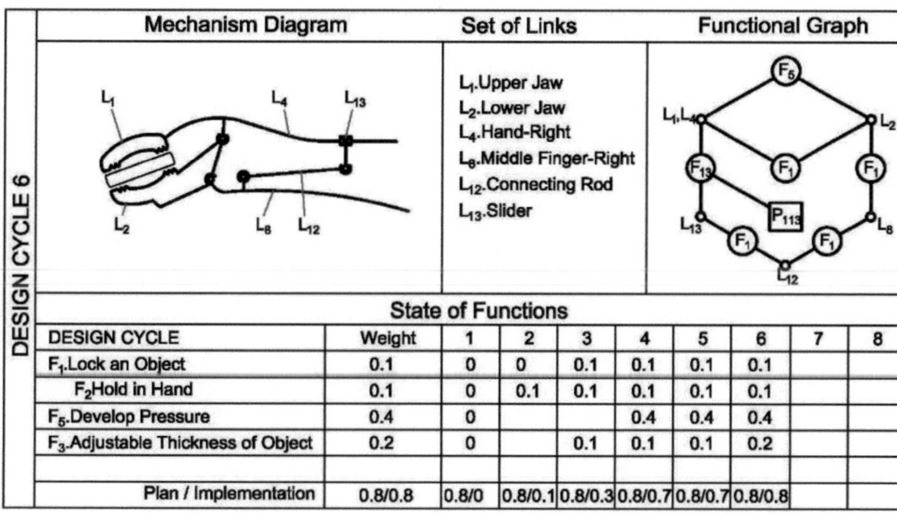

State of Functions

DESIGN CYCLE	Weight	1	2	3	4	5	6	7	8
F_1.Lock an Object	0.1	0	0	0.1	0.1	0.1	0.1		
F_2Hold in Hand	0.1	0	0.1	0.1	0.1	0.1	0.1		
F_5.Develop Pressure	0.4	0			0.4	0.4	0.4		
F_3.Adjustable Thickness of Object	0.2	0		0.1	0.1	0.1	0.2		
Plan / Implementation	0.8/0.8	0.8/0	0.8/0.1	0.8/0.3	0.8/0.7	0.8/0.7	0.8/0.8		

Figure 15.
A slider is provided for connecting rod, allowing to achieve the adjustable feature of locking pliers.

purpose hand tool in the market, and serves as a frequently used tool for a household user. A comparison list of two products is shown in **Table 1**.

Two features of this new creative hand tool are planned and structurally implemented. Novel features of self-adjustable nut wrench are quick- and self-adjustment and one-hand operation. The disadvantage is lower level of torque ability. The adjustable nut driver (known product) has slower adjustment action, high torque, and the necessity of two-hand operation for adjustment of the distance between the jaws. It has no self-adjustment feature and human control is needed for adjustment.

The original stage of design includes situation when one of the main features of one hand operation should be revealed and described by virtual mechanisms. For that reason, the palm and three fingers of human hand namely little finger, index

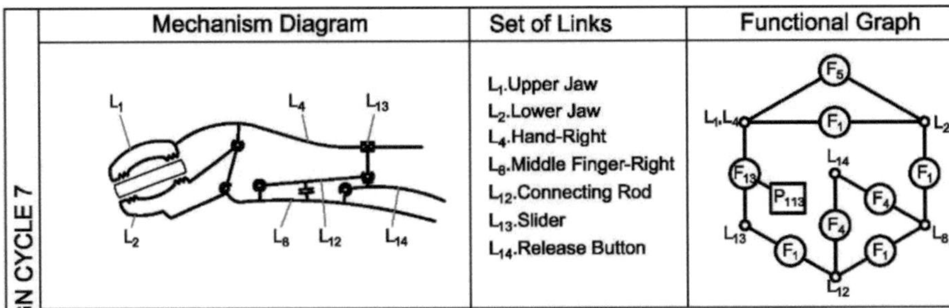

State of Functions									
DESIGN CYCLE	Weight	1	2	3	4	5	6	7	8
F_1.Lock an Object	0.1	0	0	0.1	0.1	0.1	0.1	0.1	0.1
F_2Hold in Hand	0.1	0	0.1	0.1	0.1	0.1	0.1	0.1	0.1
F_5.Develop Pressure	0.4	0			0.4	0.4	0.4	0.4	0.4
F_3.Adjustable Thickness of Object	0.2	0		0.1	0.1	0.1	0.2	0.2	0.2
F_4.Release	0.2	0	0.1	0.1	0.1	0.1	0.1	0.2	0.2
Plan / Implementation	1/0.8	1/0	1/0.2	1/0.4	1/0.8	1/0.8	1/0.9	1/1	1/1

Figure 16.
A new link L_{13} is arranged between the lower handle and connecting rod to allow the mechanism to be released from a locked state when the release button is triggered.

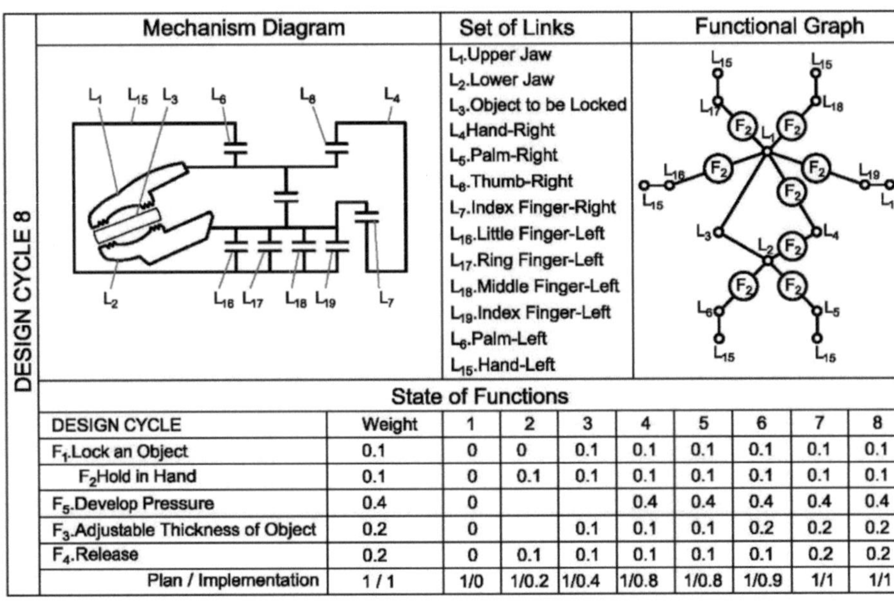

State of Functions									
DESIGN CYCLE	Weight	1	2	3	4	5	6	7	8
F_1.Lock an Object	0.1	0	0	0.1	0.1	0.1	0.1	0.1	0.1
F_2Hold in Hand	0.1	0	0.1	0.1	0.1	0.1	0.1	0.1	0.1
F_5.Develop Pressure	0.4	0			0.4	0.4	0.4	0.4	0.4
F_3.Adjustable Thickness of Object	0.2	0		0.1	0.1	0.1	0.2	0.2	0.2
F_4.Release	0.2	0	0.1	0.1	0.1	0.1	0.1	0.2	0.2
Plan / Implementation	1 / 1	1/0	1/0.2	1/0.4	1/0.8	1/0.8	1/0.9	1/1	1/1

Figure 17.
This design cycle is revealing the inconvenience of using the locking pliers for release function; so far two hands are needed for this operation.

finger and palm of human hand are described by granting one-degree movements duplicated by drive chains each of them attached to the future handle of future tool to implement four functions: holding, adjustment, screwing, and releasing operations by a single hand. So the model at this stage of conceptual design includes human hand, handle of the tool, nut, bolt, ground floor, human feet, human body, and the chain, which is closed at the already mentioned human hand.

The next set of functions lying in the model for action synthesis of the mechanism responsible for fast, not reversible, and secure adjustment holding the nut head during screwing operation needs to be inserted in a block located between the first set of functions. The third set of featured relates to the portioned rotation of

Adjustable nut driver: known product	Self-adjustable nut wrench: new product
Two-hand usage	One-hand usage
Not so convenient	Convenient
Adjustable	Adjustable
Slow adjustable	Quick adjustable
Heavy duty	Light duty
Ratcheting	Ratcheting
Higher torque	Lower torque
Not convenient to release	Convenient to release
Not easy ratchet adjustment	Easy ratchet adjustment
Range of adjustment: wide	Range of adjustment: narrow
Two-hand adjustment is needed	Self-adjustable

Table 1.
Feature list of existing and new nut wrenches.

screw or the nut head that leads to the accomplishment of ratcheting function. Implementation of the ratcheting feature is missing in the current set of design cycles. Detailed implementation inside each set of functions is based on building of an open chain or elementary set of sliders and rotational links denoting specific functional meaning to each of them and then continued by attempts to accomplish those movements by the synthesis tools of grating degrees of freedom, by freezing of movements, duplicating by parallel chains, and conditioning by drives.

4.2 Ten sets of design cycles for conceptual design of self-adjustable nut wrench

Below 10 steps of conceptual design for the novel tool per predefined features are listed. Explanations for each step are cited in **Figures 18–27**.

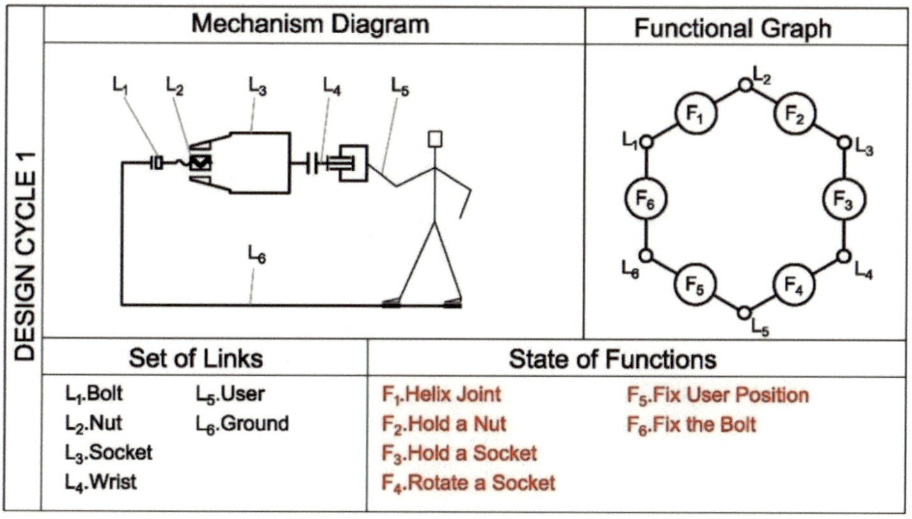

Figure 18.
The first design cycle is setting objectives of design, listed red highlighted in the section state of function. The links involved in the design are connected in a way to plan implementation of required functions.

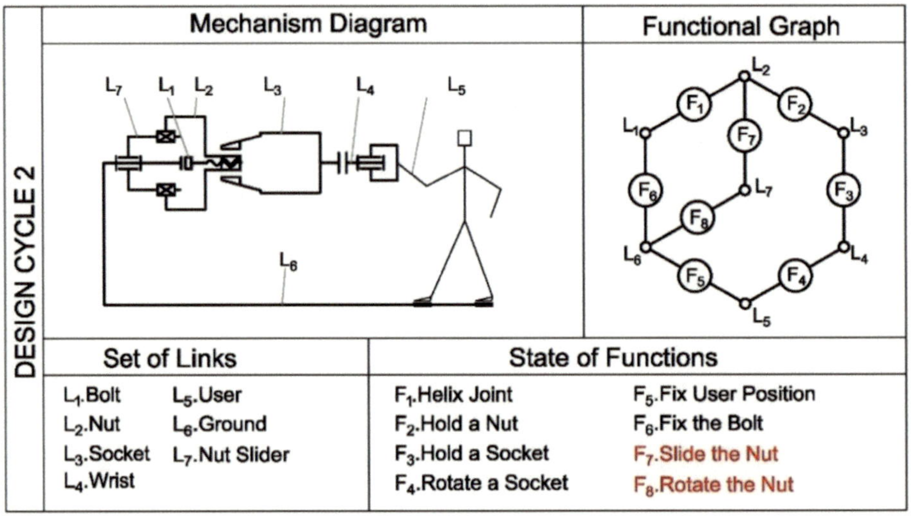

Figure 19.
The second design cycle is for modeling nut rotation operation, including necessary functions of nut rotation and axial sliding because of screw movement.

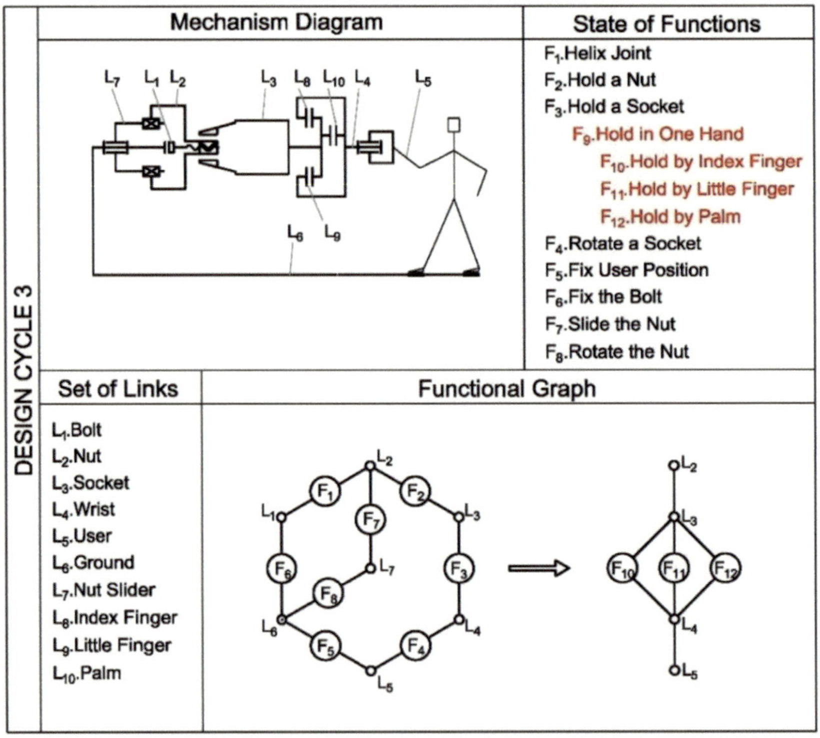

Figure 20.
The third design cycle serves for building a virtual mechanism, providing convenience of usage due to holding of the tool in one hand and providing full control during nut tightening and releasing operations.

5. Conclusions

1. The study is devoted to the development of a more visualized environment for conceptual design, based on mechanical diagrams, mechanical-functional graphs and models, and concept design result evaluation based on weighted functions.

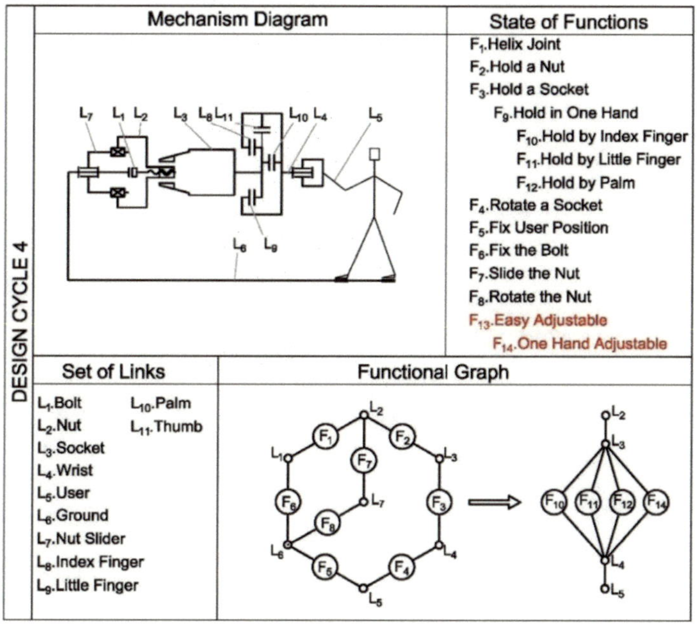

Figure 21.
The fourth design cycle is for providing convenient usage of the tool when adjusting opening between the jaws by pressing on the tool by right hand thumb while holding the body of toll by palm and two fingers of the same right hand.

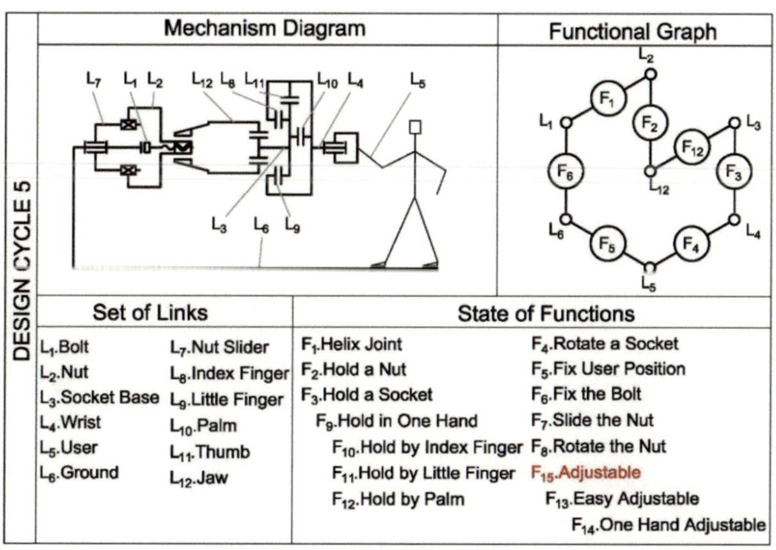

Figure 22.
The feature of adjustable size for a nut is planned in this design cycle. Jaws are separated from the socket base to allow adjustable feature. Still unknown connection between the jaws and socket body is symbolized by parallel lines urging to touch each other and belonging to neighbor links.

2. Mechanical diagrams are broken down and equipped with virtual mechanisms for detailed revelation of hidden functions for a more comprehensive consideration of functions and finding ways for their satisfaction.

3. Mechanical and functional graphs are built in a way to express both structural contents of mechanism and functions subject to implementation in both the planning state and implementation state.

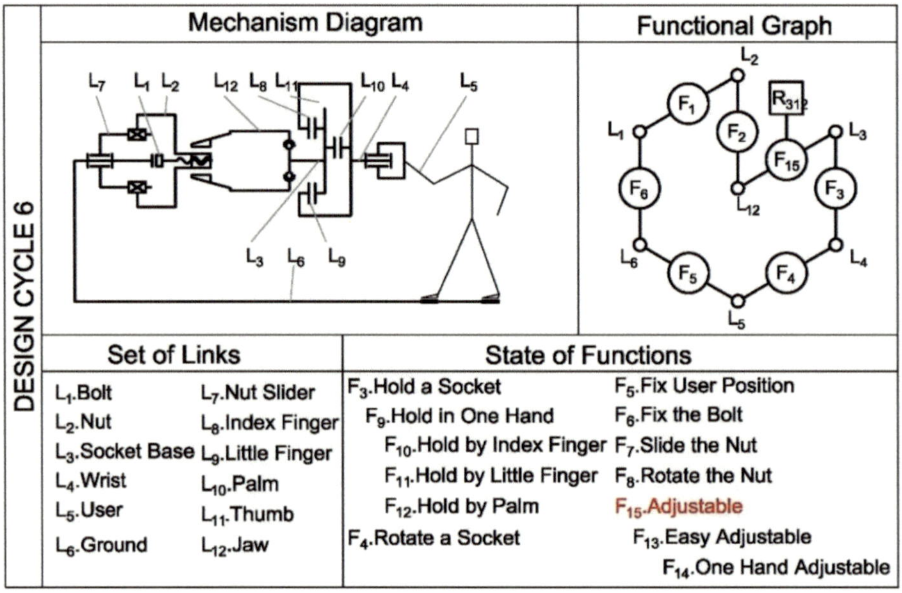

Figure 23.
The design cycle shows an implementation step of planned function (adjustable). Jaw L_{12} is linked rotationally to the socket base L_3. The graph shows this rotational connection by R_{312} symbol continued from symbol of planned function F_3.

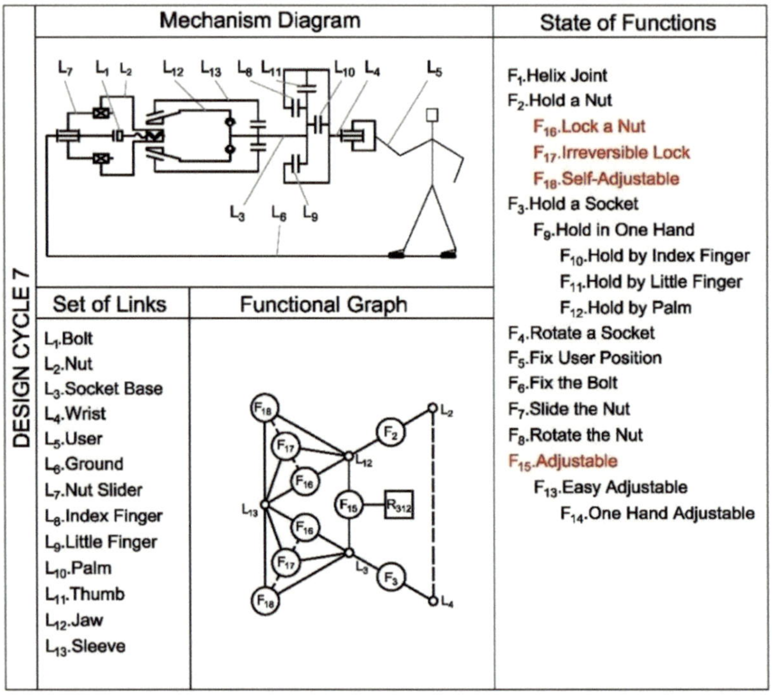

Figure 24.
The mechanical set is expanded by means of a sleeve L_{13} being in proper connection with the socket base and jaw which provides necessary functions of locking a nut, irreversible lock, self-adjustable feature, and still keeping the major function of the adjustable. This design function is a planning step for red highlighted functions.

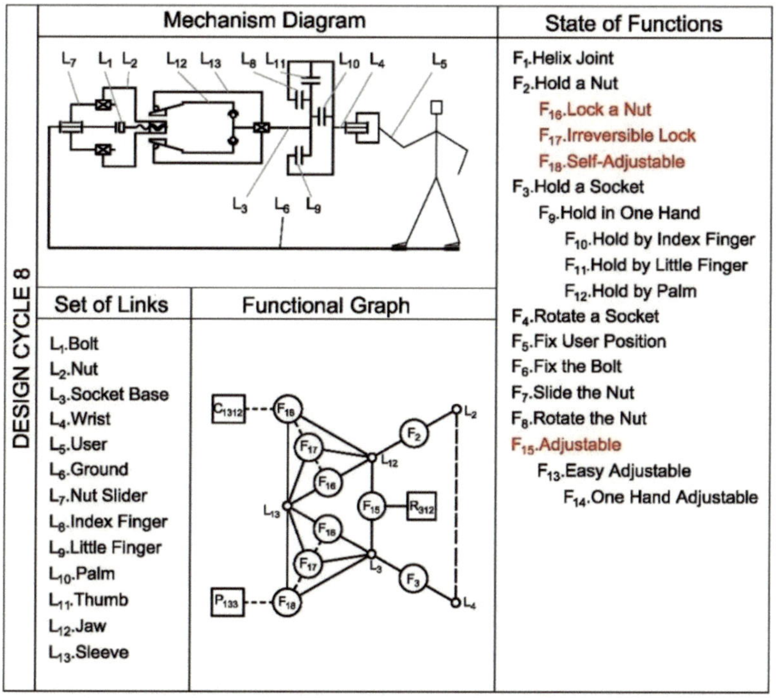

Figure 25.
At this design step, the four functions planned previously (design cycle 7) are implemented. Physical kinematical joints marked in square symbols are connecting links L_3, L_{12}, and L_{13} to provide required functions F_{15}, F_{16}, F_{17}, and F_{18}.

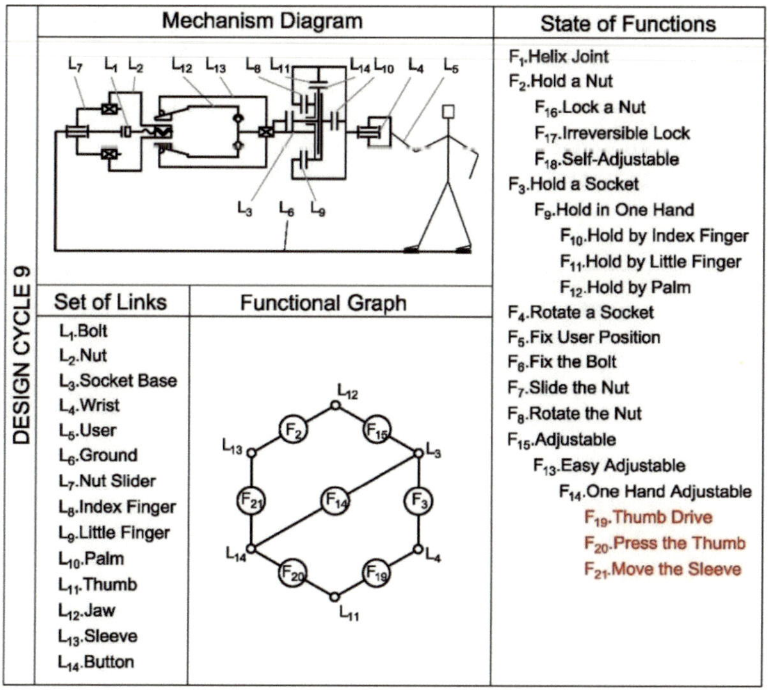

Figure 26.
The design cycle provides necessary connections between human hand finger socket base and sleeve for convenient control over jaw opening.

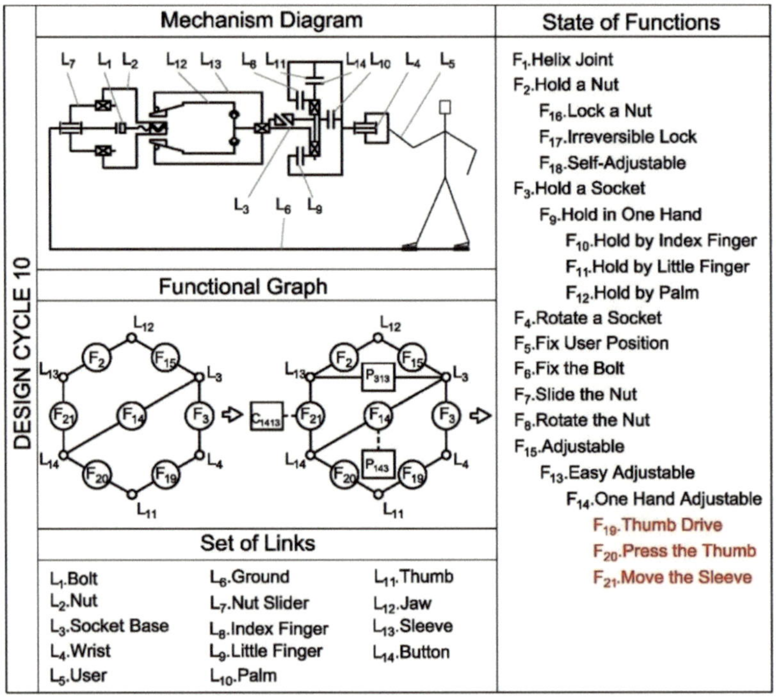

Figure 27.
Physical implementation of functions planned in design cycle 9 by duplicating functional edges of the graph with cam, rotational, and prismatic kinematical joints.

4. An exemplary set of design cycles is presented for serving needs of reverse engineering or reinventing of a known product, demonstrating, analyzing, and explaining possibilities of the method applied to a known product.

5. A hierarchical set of the traditional presentation of functions and its evaluation is upgraded into a matrix formatted table, allowing to track step-by-step implementation of tasks, still keeping hierarchical relations and numerical evaluation of conceptual design tasks at each step of the design cycle.

6. The same visualized approach is applied for structural synthesis of a novel product based on novel advantageous features and based on property comparison with a competitive product.

Author details

Hrayr Darbinyan
Shanghai Kunjek Handtools and Hardware Co. Ltd., Shanghai, China

*Address all correspondence to: hvdarbin@yahoo.com

References

[1] Chiou S-J, Kota S. Automated conceptual design of mechanisms. Mechanism and Machine Theory. 1999; **34**:467-496

[2] Paredis CJJ, Khosla PK. Synthesis methodology for task based reconfiguration of modular manipulator systems. In: Proceedings of 6th International Sumposium on Robotic Research, Hidden Valley, PA. 1993

[3] Arlitt R, Van Bossuyt DL, Stone RB, Tumer IY. The function-based design for sustainability method. Journal of Mechanical Design. 2017;**139**: 041102-041103

[4] Freudenstein F, Maki ER. The creation of mechanisms according to kinematic structure and function. Environment and Planning B. 1979;**6**: 375-391

[5] Hong-Sen Y. A methodology for creative mechanism design. Mechanism and Machine Theory. 1992;**27**(3): 235-242

[6] Pahl G, Beitz W. Engineering Design, A Systematic Approach. NY: Springer; 1995

[7] Altshuller GS. Creativity as an exact science. In: Theory of Solving Inventive Problems. Moscow: Sov. Radio, Cybernetics; 1979

[8] Moehrle MG. How Combinations of TRIZ Tools are Used in Companies-Result of a Cluster Analyses. Blockwell Publishing Ltd, R&D Management; 2005;**35**(3):285-296

[9] Darbinyan HV. Task based conceptual design method. In: Proceedings of 13th World Congress in Mechanism and Machine Science. Mexico: Guanajuato; 2011. pp. A23-A559

[10] Darbinyan HV. US Patent 8,485,070. T-Handle Wrench. 2013

Chapter 2

Integrated Model of Product Design Methods

Rosnani Ginting

Abstract

A critical factor in product innovation creativity is the development of design methodologies in various fields. The design and manufacture of a product, whether new or existing, is a significant part of engineering activities. The ability to design, develop, and produce products that customers want efficiently is the key to success in today's dynamic global market. Among these capabilities is the ability to design products that are competitive, cost-effective, and ready to be marketed on time. One key factor for maintaining competitiveness in the market is the focus on product and innovation processes by using various integrated design methods that are implemented as a standard part of design activities. The innovative integrated method, which combines various product design methods precisely can solve the main contradictory problems in the process from product demand analysis, product design, to production.

Keywords: product design, integrated model, quality function deployment (QFD)

1. Introduction

In the actual market situation, manufacturing companies must develop products that can be accepted by the customer, and at the same time, this product must be able to give satisfaction to the customer. Product design must be optimized by considering the costs, design requirements, and customers' needs [1]. The economic success of a manufacturing company depends on the ability to identify customer needs, then create products that meet those needs at a low cost. The company strategy is strongly related to the design and production processes with the most optimum level of the product quality based on customers' needs [2].

Various priorities can be increased competition in the company: quality, the speed of delivery, cost, innovation, and product limitations [3]. The customers' satisfaction and optimizing the total value of the product design is an essential goal for product development time. After defining the design of the product, the production cost can be used to create a new alternative of product design.

Many companies have tried various new approaches in product design to stay competitive. With globalization, enterprises have to compete with both local and international companies. Many methods and techniques are used by some manufacturing companies to enhance product competitiveness by fulfilling customer desire and satisfaction by improving the quality of product design. Many researchers suggest a variety of design tools that were implemented early in the design process. Therefore, various design techniques have already been developed, generated, and some of them are implemented as a design activity in some manufactures [4]. Various methods have been developed to help collect, organize, analyze, synthesize, and display the information used in the design process [5]. According to Sakao [4], various design guidelines have been developed, while a large number of individual design methods and tools have been generated, of which some were implemented as a standard part of design activities.

2. Design on engineering perspective

Design based on an engineering perspective is the application of scientific, mathematical, and creative concepts that are imagined into structures, machines, and systems that display the functions of an engineering perspective. In the process of designing consumer products in addition to the form and function of the product, engineering and industrial design are very important in the development of these products depends on the engineer and industrial designer, where the engineer functions as a determinant of the product function and the industrial designer functions to add aesthetic value in the design.

The company's ability to design, develop, and produce products that customers want efficiently is the key to success in today's dynamic global market. Among these capabilities is the ability to design products that meet the demands of competitive and cost-effective products and are ready to be marketed on time. So, companies need to develop strategic goals based on achievement, which creates a competitive advantage in the market. However, these efforts often have limitations in establishing a systematic and consistent set of methods.

3. Product design & development

Product design is the process of developing practical and effective ideas for producing new products, encompassing all the engineering and industrial design work used to develop products, from initial concept to production [6]. In this phase, important decisions are made that affect other activities. The general product development goal has not changed much over time: design a product that sale lots of the right margin. Another way to say this is: design the right product the first time, while designing the product right the first time. Product design has been widely studied in order to create methodologies that are generic enough to develop new products. The systematic method developed by Ulrich and Eppinger [7] structures the product development process according to four stages (see **Figure 1**).

This product-oriented approach defines the design process as a sequence of different phases. The transition from the task to the solution takes place in a succession of different stages. Many academic practitioners and researchers have proposed many design principles and methods to improve the quality of design, and some design methods are implemented as part of the design activities of some manufacturing companies. Each of these phases makes it possible to detail a result. Thus the

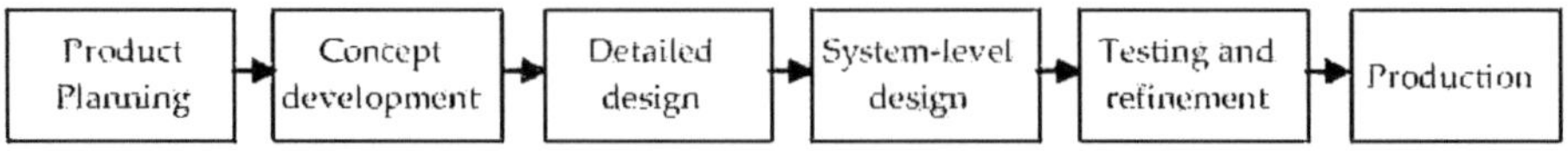

Figure 1.
Product development process [7].

specifications of the design problem are determined when the issue is clarified and defined. These specifications are then used during the various stages of the development process. Then comes the definition of the structure of the functions, the principles of solution, the structural model, technical plans and then production documents.

4. Integrated of product design methodology

The ability to quickly identify the required method is an essential priority in product achievement and design process [8]. Any company that meets the specifications and requirements of the customer will usually be more competitive than the others [9]. The ability of a company to efficiently design, develop, and manufacture customers' favorite products is key to its success in today's dynamic global marketplace. Among these capabilities is the ability to design products that meet customer needs at a competitive cost and are ready to be marketed promptly [10]. So, companies need to develop strategic goals based on achievements that create a competitive advantage in the market [11]. However, these efforts often have limitations in establishing a systematic and consistent set of methods.

There are many cases where it is difficult to find a product by merely relying on today's technology, such as technical innovation, and the identification of available technological collections in other sectors or areas will be a crucial factor. Therefore, a company must be able to innovate to meet customer needs [12]. A key factor in product innovation creativity methods is the development of design methodology methodologies that have been developed for product development [10, 13–15].

Lance and Bonollo [16] argued that the design method was about procedures, engineering, and design process. The development of design methodology includes research on design principles, practices, and procedures. The main focus on developing a product requires a deep and practical understanding of the design process and how it can be modified to be more productive.

Many academic researchers have proposed various principles of design ethics to improve the quality of design, some of which have been implemented as part of design activities in some manufacturing companies [4]. As a paradigm for simultaneous engineering design processes, it is possible to adopt various design theories and methodologies commonly used in designing a product [17].

4.1 Quality function deployment method

QFD has been recognized as an effective method for integrated products. QFD is a structured approach for integrating customer voice into product design and development [18]. The introduction of the QFD into the Americas and the European Region began in 1983 and today, and QFD continues to provide strong inspiration worldwide in the academic and manufacturing world. It is widely used in many industries such as the automotive, electronics, construction, and services sectors [19]. QFD is a multifaceted process, offering the greatest potential for significant benefits [20]. QFD is recognized as an effective method for the development of integrated processes and products [21]. The QFD aims to increase customer satisfaction based on their needs and also to enhance the profitability of the company [22]. In other words, QFD is a way of transforming the customers' desire into product design [23]. Further, Lai et al. [24], stated that QFD is a general concept that provides a means for translating customer requirements into technical requirements.

QFD is a systematic approach that determines consumer demands or requests and then translates these demands accurately into technical, manufacturing, and appropriate production planning. Revelle [25] argued that QFD was created to help an organization improve its ability to understand its customers' needs as well as to respond to those needs effectively. It means that QFD is created to help organizations improve the organizational capacity to understand customer needs, and respond effectively. QFD method is used because it can identify the customer needs and provide solutions to the existing problems. QFD described by house of quality contributes to the company about the attributes that need to be prioritized, improved and meet the customer needs.

Bouchereau and Rowlands [26], argued that the starting point of the QFD is customer preference, though often cited but measurable. These requirements will then be converted to technical specifications. Each phase of the QFD matrix represents a more specific aspect. However, only one of the essential aspects is moved into the next matrix.

Various names know the QFD matrix; the most common is the quality house (HoQ). HoQ introduces cross-linking between customer needs and design change and between the design variants themselves. Each customers' requirement is converted into one or more technical specification at all levels of the structured project with interrelated matrix [24, 27, 28] (see **Figure 2**).

QFD is a method for developing design quality that aims at customer satisfaction and then translates these needs into design goals and quality assurance points to be used at all stages of production. QFD has been recognized as an effective method for integrated products. QFD is a structured approach to integrate customer voices into product design and development [18]. QFD continues to provide strong inspiration in the academic and manufacturing worlds [19]. QFD is recognized as an effective guide to the development of integrated processes and products [21]. The objective of QFD is to increase customer satisfaction of product fulfillment requirements and to increase the company's profit [22].

In other words, QFD is the method to change the customers' needs in product design [23]. Furthermore, Lai et al. [24] argued that QFD is a general concept that provides methods for translating customer needs into technical specifications. QFD is implemented as a multi-phase process, offering the greatest potential to realize significant benefits [20]. Bouchereau and Rowlands [26], also argued that the starting point of QFD is the customers' wishes, although often referred to but measurable. These needs will then be changed to the technical specification. Each QFD matrix phase represents a more specific aspect. However, only one of the essential aspect is deployed into the next matrix.

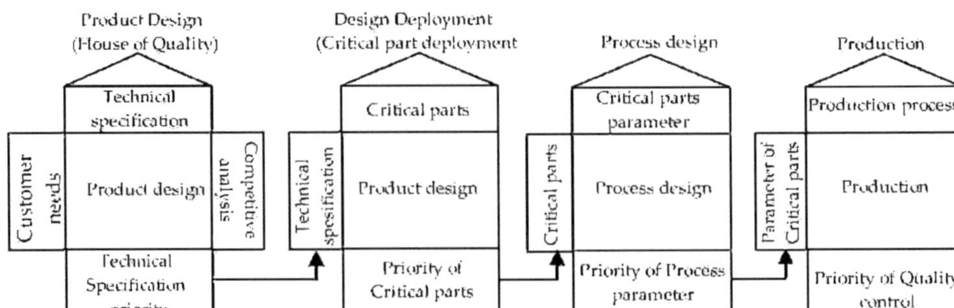

Figure 2.
QFD Matrix [29].

4.2 QFD: it's advantages and drawbacks

Generally, QFD facilitates the organization in (1) understanding the needs, (2) prioritizing customer needs, (3) communicating between team experts to ensure decision making and reducing data loss, (4) designing a product that meets or exceeding customer requirements, and (5) strategic product. Hales and Staley [30] stated that using QFD can produce better product development at a cost paid by the customer. Also, based on its customer in different company, the benefits and the advantages of some of the research done, such as customer satisfaction and reduced product production time [31], improved communication through teamwork [32] and better design [33]. Also, Bicknell in Chan and Wu [34] reported that significant benefits when QFD were used a 30–50% reduction in engineering change, a 30–50% shorter design cycle, 20–60% lower startup cost, and a security claim 20–50% less.

Lai et al. [24] recognized that QFD has great benefits that can help companies provide a better product, enhance their competitiveness in the market, and increase customer satisfaction. Poel [35] showed that the main objective of QFD is to translate customers' wishes as a goal for product specification. However, QFD is not always easy to implement, and some companies have problems using it, especially in large numbers, as well as complex systems. The problem is conventional QFD is not optimal because it's every stage of the process is subjective and qualitative during data collection to meet customer desires and to obtain technical specifications and critical parts when conducting analysis.

On the other hand, the various problems faced at various stages of QFD implementation have been widely reported in the study, in particular the traditional QFD method [36, 37]. First, the methodological framework of the conventional QFD method is no longer suitable to meet the design and product development requirements [4, 34, 38, 39]. Second, The QFD matrix is too large, and Third, the time required for the matrix sequence deployment is too long, and the product time to be marketed is not acceptable [8]. Then fourth, QFD is difficult to meet the needs of different customer groups or segments [40]. Fifth, the customers' voice is still qualitative, cannot be measured, and often misleading, it is not systematic, and the terms of product function too complicated (in this case, engineering process) are not easily determined [36]. Sixth, the customers' requirement translated into engineering terms (technical specifications) obtained from the company is still vague, too subjective, difficult to verify, and expressed in the linguistic form [40, 41]. These problems or drawbacks prompted the need for other approaches to be added when applying the QFD method. There are many different methods for generating new ideas and selecting the ideas to create a new design or to improve existing ones. Combining QFD with other techniques helps to address these drawbacks and can form the basis of future research. The integrated innovation method, which combines QFD with another technique tool, can precisely solve main contradictory problems in the process from the product demand analysis to the product design, production, and application.

4.3 An integrated model of QFD and TRIZ for product design

Although QFD has many advantages, there are still some general implementation problems in QFD. Many investigators have shown that the first phase and the second phase of QFD have many specific limitations and need to be combined with other technical norms. Kazemzadeh and Behzadian [42], have analyzed 650 articles about QFD and grouped them into four broad categories, namely general introduction, functional areas, refinery applications, and literature development.

Their findings show that some of the limitations of QFD, which need to be combined with particular applications to break QFD restraints.

QFD not only deals with product functions but also quality specifications. QFD can be accomplished by considering the adverse effects and evaluating the repair options. The TRIZ methodology can support better designers to find improvement solutions. Therefore, it is used in conjunction with QFD, because TRIZ methods, based on integrated innovation methods, can be organized in many ways. An essential element of TRIZ is conflict [43]. The essential aspects of TRIZ are discrepancies, 40 principles of creation, matrices, and scientific implications [21]. Also, the design of discrepancy matrices is useful for detecting the adverse effects of technical specifications under other improvements [44].

The synergy achieved between the four phases of QFD and TRIZ is a powerful tool for enabling product development in improvement as it emphasizes error prevention practices [10]. The synergies achieved can detect issues such as characteristic conflicts in goal specification as well as negative interactions between product structure, materials, manufacturing processes, and production control specifications.

Many researchers have worked on QFD and TRIZ combination and deployed TRIZ to address QFD problems and shortcomings. For example, Wang et al. [44] identified contradictions within TRIZ by defining methods based on HOQ (House of Quality) in QFD. Various main parameters can be extracted and used to resolve conflicts and contradictions in QFD [45]. Regazzoni et al. [46] pointed out that taking an innovative, active, and prospective approach is much more effective than showing passive reactions in preventing product collapse during its initial designation stages. TRIZ instrument was implemented to resolve these conflicts by translating the technical requirements into 39 designation parameters.

In the contradiction matrix, ameliorating parameters in rows and deteriorating parameters are arranged in columns. As QFD reveals the "what's" of required operations, the TRIZ instrument determines the "how's" of the required procedures [5]. Sakao [4] presented TRIZ as a set of technology trends related more to quality control. The purpose is to help designers to become more efficient in making improvements changes to their designs. The designers need only to focus on more

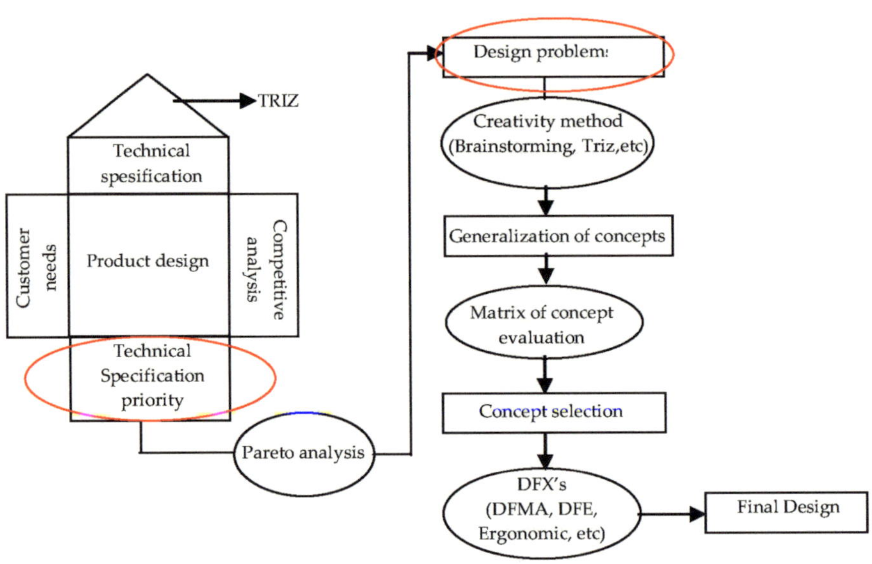

Figure 3.
QFD and it's application [5].

influential components to improve the quality aspect of a product. This is because QFD reveals "what" of the required operations, while TRIZ instrument determines "how" of the required operations. Farsijani et al. [5] addressed the combination of QFD and TRIZ as in **Figure 3**.

The TRIZ method focuses on solving problems and constraints inherent in QFD. For example, Wang et al. [44] identified contradictions in TRIZ by defining quality-based home methods in QFD. Regazzoni et al. [46] showed that innovative approaches, more active and prospective. The correlation matrix at the top of the quality house is the key QFD integrated with TRIZ [47]. **Table 1** shows a list of some relevant literature on sectors and areas related to TRIZ integration with QFD [27–33].

As shown in **Table 1**, various previous studies have integrated the TRIZ model into the QFD process, mainly in defining the optimum technical specification priority in product design. Numerous previous studies have systematically integrated QFD with TRIZ and enabled effective and systematic technical innovation for new products. TRIZ was developed to help engineers find innovative solutions during the technical product development process. Some case studies show that the proposed model enables designers to find solutions that are simple, innovative, and customer-centric. Therefore, researchers conclude that TRIZ can help QFD quantitatively identify technical requirements and critical section with inventive principles 40 and 39 parameters.

Year	References	Variables	Applied in
2002	Yamashina et al. [14]	Customer requirements and technical specifications	Design of washer
2004	Marsot and Claudon [48]	Customer requirements and technical specifications	Knife design
2009	Rau and Fang [49]	Technical specifications and product changes	Computer packaging design
2010	David et al. [50]	Customer requirements and technical specifications	Design a message change tool
2010	Butdee [51]	Customer requirements and technical specifications	Design of high-performance machines
2011	Yeh et al. [10]	Customer requirements and technical specifications	Design a laptop computer
2012	Yihong et al. [6]	Product details	Material design/construction
2013	Melgoza et al. [52]	Customer requirements and technical specifications	Design of biomedical equipment (stent tracheal)
2013	Shihdan Chen [53]	Product specifications, product details and costs	Design of mobile health tools
2013	Farsijani and Torabdaneh [5]	Customer requirements and technical specifications	Design of power transformers
2016	Patel and Deshpande [54]	Customer requirements and technical specifications	Total performance excellence design
2016	Dos Santos et al. [55]	Customer requirements and technical specifications	Design of lego foam toys
2016	Suzianti et al. [56]	Customer requirements and technical specifications	Laptop computer design

Table 1.
TRIZ method of solving technical conflicts.

4.4 A novel of QFD combined TRIZ methodology

At this phase, the identification of technical specifications and important parts objectively uses the technique of Brainstorming. The next step is to design the proposed concept of an appropriate integration method in dealing with the times that occur in the QFD process of the first and second phases, especially in addressing the contradiction between the technical specification variable (phase one of QFD) and the critical part variable (phase two QFD). This integration built through a combination of the QFD framework with the TRIZ method in a systematic and more integrated manner. Stages of the proposed Integrated QFD (IQFD)-TRIZ methodology framework can be seen in **Figure 4**.

The proposed QFD-TRIZ integration methodology in **Figure 5** which is used to develop the QFD model combined with a more integrated TRIZ can be described in fourteen stages in the followings:

1. To determine and establish research objects.

2. To identify product variables that will be used as question items for the questionnaire.

3. To identify customers' complaints through a questionnaire survey to obtain information related to what customers complained.

4. To identify customers' desires by distributing open questionnaires based on information obtained from customers' complaint questionnaires and to identify customers' needs.

In this phase, customers need to identify product variables uses a questionnaire with a Likert scale.

5. Test the content validity and reliability of the questionnaire through the criterion related validity test, and test reliability with Alpha Cronbachs' coefficient test technique.

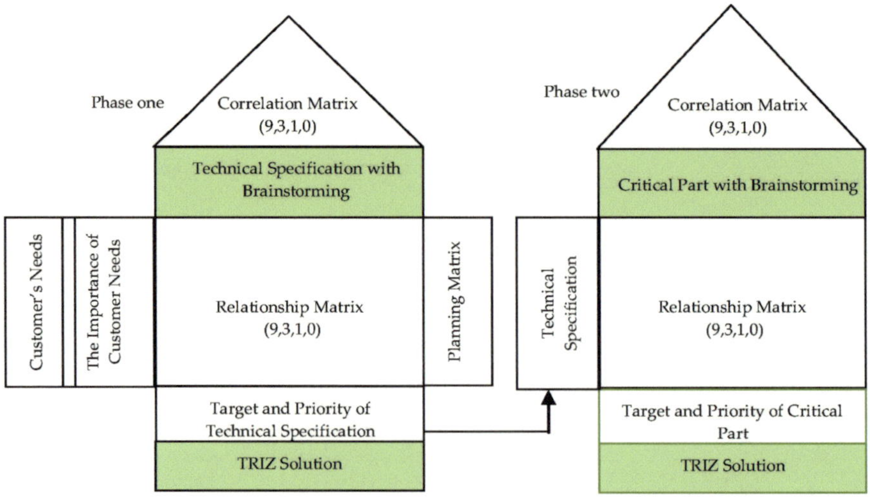

Figure 4.
Novel Integration model of QFD-TRIZ.

6. Establish relationships among technical specifications (correlation matrix) uses a questionnaire with weight 9,3,1,0

7. Determine the phase of the relationship between customer requirements and technical specifications, uses the survey, submitted to the company, weighing 9,3,1,0

8. Define and define the planning matrix

9. Define technical specification priority and targets

10. Design improvements to phase one QFD uses TRIZ methods

11. Determination of critical parts is done with the Alerting method

12. Establishing relationships among critical sections (correlation matrix) by analyzing weights 9,3,1,0

13. Determine the phase of the relationship between the Critical Divisions in phase two. The calculation is done by analyzing 9,3,1,0 weight

14. Design improvements to QFD phase two use TRIZ methods

15. Product design.

Meanwhile, the existing integration methodology of QFD-TRIZ model by Melgoza et al. [52], was used as the development of a Novel integration model developed in the study (see **Figure 5**). The QFD integration methodology steps with the TRIZ model have been selected to be compared with the Novel IQFD-TRIZ. The stages of the existing model as follows:

1. Identification of customer requirements matrix

In this phase, customer voice recognition to know what customers want in terms of product design is done through the product survey. The data are qualitative data.

2. Engineer's voice matrix identification

In this phase, defining the technical specifications of the existing IQFD – TRIZ method is done uses Interviews.

3. Identification of the interest phase of the customer change

This phase uses a questionnaire with the Likert scale.

4. Determine the phase relationship matrix

In this phase, the relationship between customer requirements and technical specifications uses the 5,4,3,2,1 scale.

5. Determine the phase correlation matrix

6. In this phase, the relationship between each technical specification uses the 9,3,1 scale.

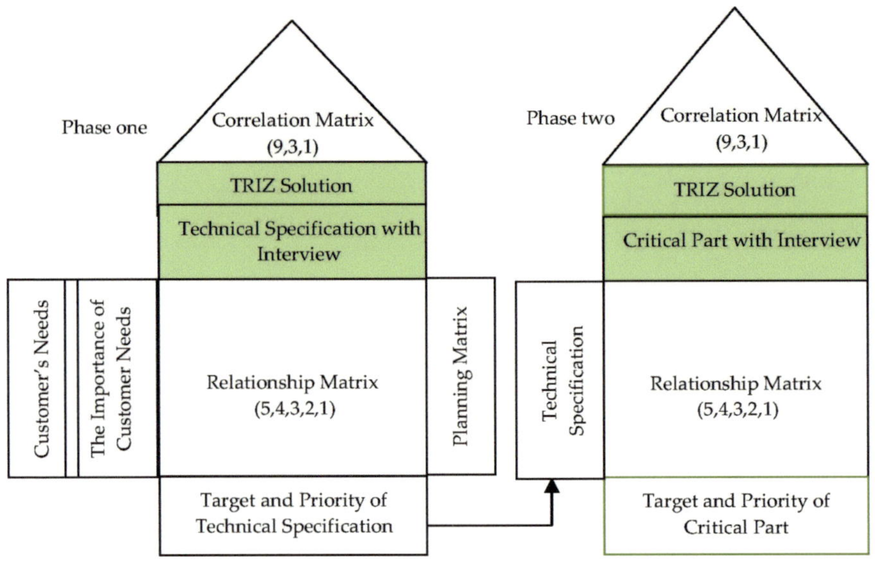

Figure 5.
Matrix of QFD-TRIZ model by Melgoza et al. [52].

4.4.1 Discussion of how to use the novel QFD framework compared to existing QFD

The main focus of this research is to optimize the performance of the QFD process integrated with quantitative design techniques, namely comparing novel QFD integration components (novel IQFD) with existing QFD integration (IQFD). The combination built into this research is compared to the existing one developed by previous research. The components of the novel IQFD methodology framework developed in this research maximize product design models efficiently, structurally, quantitatively, systematically, and propose new solutions in designing new products and company products maximizes the functionality of the TRIZ model into the QFD process.

Meanwhile, the weights and scales used in the novel IQFD use the 9,3,1.0 scale much better than the existing IQFD uses the + 2, +1, 0 −1, −2, 4, 3, 2, 1 scale, 0 and scale 5, 4, 3, 2, 1.0. Besides, to identify customer requirements, technical specifications and critical parts of the novel QFD component use the "Brainstorming" method, and the resulting changes are more relevant than the existing QFD use "Interview" method that performed by only one person in the company. The contribution is used to stimulate open discussion of product creative ideas and improvements made to various sources of information, namely, research, specialist, marketing, sales, production, and management.

4.4.2 Discussion the difference of novel QFD framework compared to existing QFD

Some previous studies have discussed QFD integration with various quantitative design techniques models to overcome the constraints of the QFD process in each of its phases. However, there are still limitations in its implementation. This section will discuss in stages how to use the novel IQFD framework component in comparison to the existing IQFD conducted by previous research.

How to use the novel IQFD-TRIZ framework built in this research in comparison to the existing IQFD-TRIZ developed by Melgoza et al. [52], can be seen in **Tables 2** and **3**, as follows.

Item	Novel IQFD-TRIZ	Existing IQFD-TRIZ
Customer requirements	Conventional QFD	Conventional QFD
Technical specifications	The burden of suffering	Interview method
Relationship matrix	Weight 9,3,1,0	Weight 5,4,3,2,1
Correlation matrix	Weight 9,3,1,0	Weight 9,3,1
Triz solution	Emphasized after analysis of target matrix and technical specification priority	Emphasized after correlation matrix analysis
Extra space/matrix	1. Technical specification contribution 2. Triz solution	Triz solution

Table 2.
The differences in the novel IQFD-Triz analysis vs the existing IQFD-Triz (Phase One).

Table 2 above can be described as follows:

1. Identifying customer requirements of the Novel IQFD-TRIZ and IQFD-TRIZ models uses questionnaires to determine product design changes. After identifying the customers' needs, it defines the importance of the customers' requirements. Novel IQFD-TRIZ built on this research, and the existing IQFD-TRIZ developed by Melgoza et al. [52] uses a Likert scale

2. Identification of technical specifications (phase one), and weighting of interests.

   After identifying customer requirements, technical specification is implemented with Brainstorming method used in the proposed IQFD-TRIZ, while the existing IQFD-TRIZ uses the interview method.
   As discussed earlier, it is found that customers' voice is still qualitative, that is, measurable and often misleading, systematic. At the same time, the product function requirements are too complex and, therefore, not easy to determine. Changing customer requirements translated into technical specifications obtained from companies is still unclear, too subjective, difficult to verify and only expressed in linguistic form, so the study emphasized that the problem is solved mainly in phase one and phase two, then the novel IQFD framework component built in this research uses Brainstorming method, compared to the existing IQFD that uses the interview method.
   Brainstorming methods focus on companies, from top management to middle management, are involved in defining technical specifications and critical areas, making these changes more objective than those of the incorporation rules. Existing QFDs, where technical specifications and critical parts are made to one person from the company, the production manager, make the changes more subjective. Through the use of Brainstorming, it is possible to obtain more specific technical and critical properties than just one person, the production division manager. The focus of Brainstorming in defining technical specifications and critical parts of the novel QFD framework components is to analyze the translation of customer requirements into technical specifications in phase one and also technical specifications into critical parts in phase two objectively, efficient and accurate translation. Also, the identification of technical specifications and critical parts with Brainstorming can identify product changes that are in line with customer requirements.

3. Define the relationship (relationship matrix) between customer requirements and technical specifications (phase one).

   Subsequently, defining the relationship between customer requirements and technical specifications, in phase one, and technical specification relationships with critical parts in phase two were performed on the Novel IQFD-TRIZ uses 9,3,1.0. Meanwhile, the existing IQFD-TRIZ model developed by Melgoza et al. [52] uses a scale of 5,4,3,2,1.

4. Define the correlation matrix between technical specifications (phase one).

   Correlations between technical specifications in phase one of the novel IQFD-TRIZ model were performed uses the 9,3,1.0 scale, whereas the existing IQFD-TRIZ model uses the 9,3,1 scale.

5. Resolving conflicting differences between technical specifications.

   In this phase, the TRIZ method was applied to resolve conflicting issues that occur in phase one QFD. Conflict resolution was performed in various stages, namely (1) defining specific problems, (2) defining common problems uses a 39 × 39 matrix conflict table, (3) identifying joint solving uses 40 TRIZ rules.

6. Define planning matrix (phase one).

   After resolving the conflicts in the correlation matrix, then the novel QFD-TRIZ and the existing perform an analysis of customer requirements and the technical specification to prevent design changes in the next phase. In phase two of the QFD, by completing the calculation and determination of the matrix planning, which defines the planned weight loss, calculates the ratio of improvement value to ratio, absolute weight, and relative weight.

7. Define target and technical specification priority (phase one).

   The goals and priorities of the critical sections of the novel IQFD-TRIZ and the existing IQFD-TRIZ are both defined to assess the importance of which critical parts are of the highest weight, the difficulty level in designing the smallest product, and the lowest design cost.

Meanwhile, for the differences between the novel IQFD-TRIZ compared to the two-phase IQFD-TRIZ can be seen in **Table 3** as follows:

Item	Novel IQFD-TRIZ	Existing IQFD-TRIZ
Critical part	Brainstorming method	Interview method
Relationship matrix	Weight 9,3,1,0	Weight 5,4,3,2,1
Correlation matrix	Weight 9,3,1,0	Weight 9,3,1
Triz solution	Emphasized after the target matrix and critical section priority	Emphasized after correlation matrix
Extra space/matrix	1. Brainstorming on critical part	Triz solution on correlation matrix

Table 3.
The differences analysis of novel IQFD-Triz vs the existing IQFD-Triz (Phase Two).

Table 3 above can be described as follows:

1. Identification of critical parts (phase two), and weighting of interests.

   The identification of the critical part in phase two of the novel IQFD-TRIZ was made uses the Brainstorming method. In contrast, in the existing IQFD-TRIZ model developed by Melgoza et al. [52] uses the interview method. Meanwhile, the novel IQFD-TRIZ uses the Likert scale.

2. Define the relationship (relationship matrix) between technical specification and critical part (phase two).

   Defining the correlation matrix in phase two of the novel IQFD-TRIZ model uses the 9,3,1,0 scale, and the 5,4,3,2,1 scale built on the existing IQFD-TRIZ, is similar to the scale used in phase one.

3. Define correlation (correlation matrix) between critical parts (phase two).

   After obtaining the variables that are in line with the customers' needs, the next step is to perform a correlation analysis between the variables of the qualitative components defined uses the 9,3,1.0 scale in the model developed in this research. Meanwhile, in the model developed by Melgoza et al. [52] used a scale of 5,4,3,2,1.

4. Define planning matrix (phase two).

   Subsequent to defines the correlation matrix further performs customer requirements analysis and technical specifications to prevent any product design changes in the next phase, in phase two or phase three and phase four, which is done by calculating and determining the design matrix, which defines weight planned interest, calculating the ratio of the value of the improvement ratio, absolute weight, and relative weight.

5. Define the target and priority of the critical parts (phase two).

   The target matrix and the critical phase priority in phase two are defined just as they were in phase one of the novel IQFD-TRIZ or the existing IQFD-TRIZ.

4.4.3 The contribution for knowledge

Discussions on the contribution of new knowledge have been discussed through the comprehensive and phases of the QFD matrix. The proof of the contribution of this new knowledge is described as follows:

1. The design of an integrated QFD integration development framework that is oriented to customer emotional satisfaction, technical specifications, and critical parts.

2. QFD phases become more objective by facilitating the calculation of subjective matters.

3. Brainstorming method allow the idea to fill technical specifications in phase one QFD and critical phase in phase two QFD.

4. The methodology combined with QFD is the same for both phases and facilitates learning and application (the same data can be used for QFD phase one and phase two (seamless transfer)).

5. House of QFD model more integrated between phase one and phase two than existing models for each type of QFD combination.

6. House of QFD on phase one and phase two are different for each type of QFD combination where the focus aspects of product design will follow the combined technique.

5. Conclusion

QFD flexibility has facilitated integration with other engineering design tools. This research developed a framework of phase one and phase two of QFD combined Triz, in which the concept developed in this research based on previous studies. The framework components of novel QFD-Triz can effectively overcome the drawbacks while increasing QFD analysis in every phase, and also, there is a new procedure in product design. A novel of QFD framework developed in this study has the potential to be the best technique for designing quality from the customers' point of view. It is believed that this study will provide some research opportunities, for example, emphasizing on improving QFD capabilities and raising problems related to the product design.

Author details

Rosnani Ginting
Department of Industrial Engineering, Universitas Sumatera Utara, Medan, Indonesia

*Address all correspondence to: rosnani@usu.ac.id

References

[1] Prasad K, Chakraborty S. A quality function deployment-based model for cutting fluid selection. Advances in Tribology. 2016;**2016** 3978102, 10 pages

[2] Vavdhara AMK, Yadav BJS, Yadav CL, Ghosh DMK. Quality improvement in steel rolling industry through quality function deployment: A case study in Ajmera Steel Rolling, Ratlam. In: International Conference on Current Trends In Technology, 'NUiCONE. 2011. pp. 1-4

[3] Olhager J, Bengtsson J. The impact of the product mix on the value of flexibility. Omega. 2020;**30**(4):265-273

[4] Sakao T. A QFD-centred design methodology for environmentally conscious product design. International Journal of Production Research. 2013;**45**(18-19):4143-4162

[5] Farsijani H, Torabandeh MA. Improvement of efficiency in product designing by usage of fuzzy QFD & TRIZ: A case study on Transfo company. AENSI Journals. Journal of Applied Science and Agriculture. 2013;**8**(4):451-461

[6] Yihong L, Yunfei S, Ting C. Study of new wall materials design based on TRIZ integrated innovation method. Management Science and Engineering. 2012;**6**(4):15-29

[7] Ulrich KT, Eppinger SD. Product Design and Development. 4th ed. New York: McGraw Hill; 2008

[8] Prasad B. Review of QFD and related deployment techniques. Journal Manufacturing System. 1998;**17**(3):221-234

[9] Besterfield DH. Total Quality Management. 2nd ed. New Jersey, Prentice-Hall: Upper Saddle River; 1999

[10] Yeh CH, Jay CY, Huang CK, Yu. Integration of four-phase QFD and TRIZ in product R and D: A notebook case study. Research in Engineering Design. 2011;**22**(3):125-141

[11] De Felice F. A multiple choice decision analysis: An integrated QFD–AHP model for the assessment of customer needs. International Journal of Engineering, Science and Technology. 2010;**2**(9):25-38

[12] Lee-Mortimer A. Managing innovation and risk. World Class Design to Manufacture. 1995;**2**(5):38-42

[13] Li X, Tan KC, Xie M. Optimizing product design using quantitative Quality Function Deployment: a case study. Quality and Reliability Engineering International. 2007;**23**(1):45-57

[14] Yamashina H, Ito T, Kawada H. Innovative product development process by integrating QFD with TRIZ. Journal of the Japan Society for Precision Engineering. 2002;**66**(11):1705-1710

[15] Liu H-T. Product design and selection using fuzzy QFD and fuzzy MCDM approaches. Applied Mathematical Modelling. 2011;**35**(2011):482-496

[16] Lance NG, Bonollo E. The development of a suite of design methods appropriate for teaching product design. Global Journal of Engineering Education. 2002;**6**(1):45-52

[17] Gonçalves-Coelho AM, Mourão AJF, Pereira ZL. Improving the use of QFD with axiomatic design. Concurrent Engineering: Research and Applications. 2005;**13**(3):232-239

[18] Mendoza N, Horacio A, Arturo M. Case studies in the integration of QFD,

VE and DFMA during the product design stage. In: The Proceedings of the 9th International Conference of Concurrent Enterprising, Espoo, Finland. 2003. pp. 16-18

[19] Kant S, Hundal GS. Product design through QFD approach with hybrid of AHP and Fuzzy Logics. 21 International Proceedings of Economics Development and Research; 2014. pp. 112-116

[20] Naseri K. Algorithm for cost-optimized QFD decision-making problem. Evolutionary Intelligence. 2014;**9**(1-2):21-36

[21] Yang CL, Huang RH, Wei WL. A modified TRIZ for new product development management. Business and Information (Sapporo, July 3-5). 2012:323-334

[22] Gupta S, Okudan GW. Computer-aided generation of modularised conceptual designs with assembly and variety considerations. Journal of Engineering Design. 2012:**19**(6):533-551

[23] Pourhasomia MH, Khamseh AA, Ghorbanzad Y. A hybrid of Kano and QFD for ranking customers' preferences: A case study of bank melli iran. Management Science Letters. 2013;**3**:845-860

[24] Lai CJ, Hsu CH, Kuo HM. An empirical study of constructing a dynamic mining and forecasting system for the life cycle assessment-based green supply chain. Wseas Transactions On Systems. 2012;**11**(4):129-139. ISSN: 2224-2678

[25] Revelle, Jack B. The Hand Book of QFD. United States: Acid Free Paper. 1997

[26] Bouchereau V, Rowlands H. Methods and techniques to help quality function deployment (QFD). Benchmarking: An International Journal. 2000;**7**(1):8-19

[27] Kao HP, Su E, Wang B. IQFD: A blackboard-based multi-agent system for supporting concurrent engineering projects. International Journal of Production Research. 2002;**40**(5):1235-1262

[28] Hsing CH, Lo CH, Tseng KC. One-Step QFD based 3D morphological charts for concept generation of product variant design. Expert Systems with Applications. 2010;**37**:7351-7363

[29] Cohen L. Quality Function Deployment: How to make QFD work for you. Addison-Wesley. 1995

[30] Hales R, Staley D. Mix target costing, QFD for successful new products. Marketing News. 1995;**29**(1):18

[31] Hauser JR, Clausing DP. The House of Quality. Harvard Business Review, May/June. 1998

[32] Griffin A, Hauser JR. The voice of the customer. Marketing Science. 1998;**12**(1)

[33] Mehta P. Designed chip embeds user concerns. In: Electronic Engineering Times. 1994 January 24

[34] Chan LK, Wu ML. Quality function deployment: A literature review. European Journal Operation Research. 2002;**143**:463-497

[35] Poel IVD. Methodological problems in QFD and directions for future development. Research in Engineering Design. 2007;**18**(1):21-36

[36] Law HW, Hua M. Using quality function deployment in singulation process analysis. Engineering Letters. 2007;**14**:1. EL_14_1_6 (Advance online publication: 12 February 2007), pp. 1-5

[37] Mohammadi F, Sadi MK, Nateghi F, Abdullah A. A hybrid quality function deployment and cybernetic analytic

network process model for project manager selection. Journal of Civil Engineering and Management. 2014, 2014;**20**(6):795-809

[38] Akao Y, Mazur GH. The leading edge in QFD: Past, present and future. International Journal of Quality & Reliability Management. 2003;**20**(1):20-35

[39] Brad S. Complex system design technique (CSDT). International Journal of Production Research. 2004:1-28

[40] Kim KJ, Moskowitz H, Dhingra A, Evans G. Fuzzy multi-criteria models for quality function deployment. European Journal Operation Research. 2000;**121**(3):504-518

[41] Zhou M. Fuzzy logic and optimization models for implementing QFD. Computers & Industrial Engineering. 1998;**35**(1-2):237-240

[42] Kazemzadeh RB, Behzadian M, Aghdasi M, Albadvi A. Integration of marketing research techniques into house of quality and product family design. The International Journal of Advanced Manufacturing Technology. 2009;**41**(9-10):1019-1033

[43] Sarno E, Kumar V, Li W. A hybrid methodology for enhancing reliability of large systems in conceptual design and its application to the design of a multiphase flow station. Research in Engineering Design. 2005;**1-2**(16):27-41

[44] Wang H, Chen G, Lin Z. Algorithm of Integrating QFD and TRIZ for the innovative design process. International Journal Computer Application Technology. 2005;**23**(1):41-52

[45] Lu C, Liao Z, Jiang SH, Lin G. Research on Innovative Product design system based on QFD and TRIZ. China Trans Technology Publication. 2006:1144-1147

[46] Regazzoni D, Russo D. TRIZ tools to enhance risk management. TRIZ Future Conference 2010, Procedia Engineering. 2010;**9**:40-51

[47] Liu S, Shi D, Ying Z. A planning approach of engineering characteristics based on QFD-TRIZ integrated. RIFIP AICT 304. IFIP International Federation for Information Processing. 2009:117-126

[48] Marsot J, Claudon L. Design and Ergonomics. Methods for integrating ergonomics at hand tool design stage. International Journal of Occupational Safety and Ergonomics: JOSE (Impact Factor). 2004;**10**(1):13-23

[49] Rau H, Fang YT. Conflict resolution of product package design for logistics using the TRIZ method. Machine Learning and Cybernetics. International Conference. 2009;(5):2891-2896

[50] David, Novoa DR, Rovira LN, Aguayo TH, Said. Inventive problem solving based on dialectical negation, using evolutionary algorithms and TRIZ heuristics. Computers in Industry. 2010;**4**(62):437-445

[51] Butdee ST. QFDE combined with TRIZ framework to formulate and respond to functional design for a high temperature machine (htm). C. Asian International Journal of Science and Technology in Production and Manufacturing Engineering. 2010;**3**(4):77-84

[52] Melgoza EL, Serenó L, Rosell A, Ciurana J. An integrated parameterized tool for designing a customized tracheal stent. Computer-Aided Design. 2012;**44**(12):1173-1181

[53] Shih HS, Chen SH. A conceptual design of a mobile healthcare device-an application of three-stage QFD with ANP and TRIZ. International Journal of Operations Research. 2013;**10**(2): 80-91

[54] Patel A, Deshpande V. A review: quality improvement through the integration of six sigma, QFD and TRIZ in manufacturing industries. Industrial Engineering Journal. 2016;**9**(10):34-39

[55] Santos MD, Fernandes MdC, dos Santos FMC, Dias FdC., Almeida JJ, Júnior JA. An approach of triz methodology with inventive solutions for toys used by children with special needs based on the requirements of quality house (QFD). IOSR Journal of Engineering (IOSRJEN). ISSN. 2016;(12): 2278-8719. ISSN (e): 2250-3021

[56] Suzianti A, Yudha IS, Anjani S, Handana IG. Designing laptop based on costumer preference using ECQFD and TRIZ methods. 6th International Conference on Industrial Engineering and Operations Management in Kuala Lumpur. 2016;**8**(10):1132-1138

Chapter 3

Photogrammetry as an Engineering Design Tool

Ana Pilar Valerga Puerta, Rocio Aletheia Jimenez-Rodriguez, Sergio Fernandez-Vidal and Severo Raul Fernandez-Vidal

Abstract

Photogrammetry is a technique used for studying and precisely defining the shape, dimension, and position in space of any object, using mainly measurements taken over one or more photographs of that object. Today, photogrammetry is a popular science due to its ease of application, low cost, and good results. Based on these causes, it is becoming a good alternative to scanning. This has led to its implementation in different sectors such as the archeological, architectural, and topographical for application in element reconstructions, cartography, or biomechanics. This chapter presents the fundamental aspects of this technology, as well as its great possibilities of application in the engineering field.

Keywords: 3D scan, reverse engineering, 3D design, point cloud, CAD, virtual model, 3D reconstruction, virtual assembly, augmented reality, virtual reality

1. Reverse engineering

Reverse engineering is based on the study of certain principles and information of a product. The main function of reverse engineering is to obtain the maximum information about an element or device, including its geometry and appearance, among other things [1, 2]. Its first appearance was around World War II, in military operations.

The field of application of this type of engineering is very wide, highlighting the 3D digitalization used mainly for research, analysis, and reasoning of the technology used by other companies, for the development of elements without making use of specific information (redesign), and for the tasks of inspection or virtual metrology of a product in almost every industry [3].

The main 3D digitization technologies are shown in **Figure 1**, among which photogrammetry stands out for its ease of use and low cost.

2. Photogrammetry

Photogrammetry is distinguished by the measurement on photographs, allowing to obtain from any object its real dimensions, position, shape, and textures [4, 5]. These processes or this science emerged in the middle of the nineteenth century, being as old as photography. The first photogrammetric device and the first methodology were created in 1849 by the Frenchman Aimé Laussedat. He, "the father of

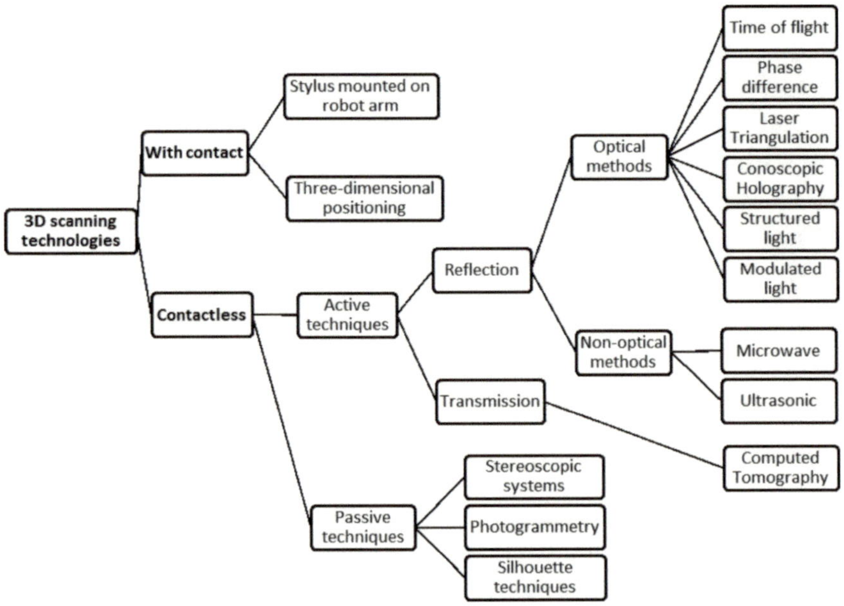

Figure 1.
Classification of 3D scanning technologies.

photogrammetry," used terrestrial photographs and compiled a topographic map. This method was known as iconometry, which means the art of finding the size of an object by measuring its image. Digital photogrammetry was born in the 1980s, having as a great innovation the use of digital images as a primary data source [6, 7].

The main phases of digital photogrammetry are analysis of the shape of the object and planning of the photos needed to be taken; calibration of the camera; image processing with specific software to generate a cloud of points; and transfer of this point cloud to the CAD software to create a 3D model. The accuracy of the reconstruction depends on the quality of the images and textures. Photogrammetry algorithms typically indicate the problem, such as minimizing the sum of the squares of a set of errors, known as "package fit" [8]. Structure algorithms, from motion (SfM), can find a set of 3D points (P), a rotation (R), and the camera position (t), given a set of images of a static scene with 2D points in correspondence, as shown in **Figure 2** [10].

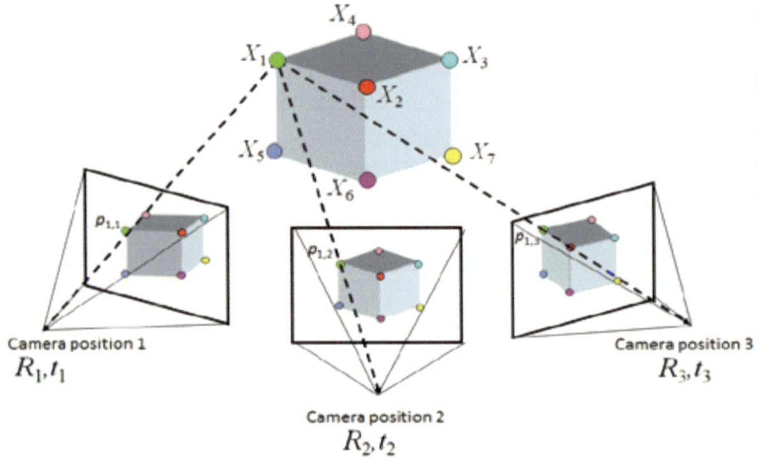

Figure 2.
Structure of the motion algorithm [9].

Photogrammetric technology is generally based on the illumination of one object and the inclusion of solutions derived from the measurement of conjugated points, appearing in two photographic images or measuring the conjunction of points in multiple photographic images (three or more images). There are different photogrammetric techniques. One of them is to ensure that the surface of the object has enough light and optical texture to allow conjugated dots to be paired through two or more images. In some cases, optical texture can be achieved by projecting a pattern over the surface of the object at the time of image capture [11–13].

3. Fundamentals of photogrammetry

The basic mathematical equations underlying photogrammetry, called collinearity equations, are responsible for unifying the coordinate system of the image in the camera with the object being photographed [14] (Eqs. (1)–(3)):

$$\begin{pmatrix} x_n - x_0 \\ y_n - y_0 \\ -c \end{pmatrix} = \lambda M \begin{pmatrix} X_n - X_0 \\ Y_n - Y_0 \\ Z_n - Z_0 \end{pmatrix} \tag{1}$$

where λ = scaling factor; M = rotation matrix; X_o, Y_o, and Z_o = the position of the perspective center in the object's space; and $p_n = (x_n, y_n)^T$ and $P_n = (X_n, Y_n, Z_n)^T$ = target n coordinates at the image plane and the space of the object, respectively. The above equation manipulated algebraically produces the well-known collinearity equations that relate the location of destination *nth* in the space of objects with the corresponding point in the plane of the image:

$$x_n - x_0 = -c\,\frac{m_{11}(X_n - X_0) + m_{12}(Y_n - Y_0) + m_{13}(Z_n - Z_0)}{m_{31}(X_n - X_0) + m_{32}(Y_n - Y_0) + m_{33}(Z_n - Z_0)} \tag{2}$$

$$y_n - y_0 = -c\,\frac{m_{21}(X_n - X_0) + m_{22}(Y_n - Y_0) + m_{23}(Z_n - Z_0)}{m_{31}(X_n - X_0) + m_{32}(Y_n - Y_0) + m_{33}(Z_n - Z_0)} \tag{3}$$

where m_{ij} $(i, j = 1, 2, 3)$ = elements of the rotation matrix M which are functions of the Euler orientation angles (ω, ϕ, κ), which are essentially the angles of tilt, rotation and rotation of the camera in the object space (Eq. (4)–(11))

$$m_{11} = \cos\phi\,\cos\kappa \tag{4}$$

$$m_{12} = \sin\omega\,\sin\phi\,\cos\kappa + \cos\omega\,\sin\kappa \tag{5}$$

$$m_{13} = -\cos\omega\,\sin\phi\,\cos\kappa + \sin\omega\,\sin\kappa \tag{6}$$

$$m_{21} = -\cos\phi\,\sin\kappa \tag{7}$$

$$m_{22} = -\sin\omega\,\sin\phi\,\sin\kappa + \cos\omega\,\cos\kappa \tag{8}$$

$$m_{23} = \cos\omega\,\sin\phi\,\sin\kappa + \sin\omega\,\cos\kappa \tag{9}$$

$$m_{31} = \sin\phi \tag{10}$$

$$m_{32} = -\sin\omega\,\cos\phi \tag{11}$$

$$m_{33} = \cos\omega\,\cos\phi \tag{12}$$

The plane of the image can be transformed analytically into its X, Y, and Z coordinates in global space. Photogrammetry is effective and computationally simple. It should be noted that its algorithm is based on definitions of both interior and exterior orientations. In a photographic system, if the internal parameters of a camera are known, any spatial point can be fixed by the intersection of two beams of light that are projected.

There are two main factors that induce photogrammetry measurement errors: System error due to lens distortion and random error due to human factors.

1. System error due to lens distortion. It causes a point in the image in the plane to move from its true position (x, y) to a disturbed position. The coordinates of any point in the image can be compensated with Eqs. (13)–(14):

$$x'_n = x_n + dx \tag{13}$$

$$y'_n = y_n + dy \tag{14}$$

In the lens, the largest error occurs at the point of the projected image. Therefore, dx, dy can be broken down by Eqs. (15)–(16):

$$dx = dx_r + dx_d \tag{15}$$

$$dy = dy_r + dy_d \tag{16}$$

2. Random error due to human factors. Theoretically, a point captured in two different photos is enough to set its 3D coordinates. To complete this, this step requires an identification and marking of the point in the two images. Any human can have failures in the marking of points, giving rise to the random error.

4. Evolution from analytical to digital photogrammetry

From the analytical photogrammetry, it is possible to describe the evolution from photogrammetry to digital, based on physical and mathematical principles. The main distinction is given by the nature of the measurement of the information taken in the images [15].

The analytical photogrammetry coordinates the image, and the gray digital image is evaluated with the digital photogrammetry. In both methods appropriate Gaussian-Markov evaluation procedures are used. Pertinent relations between object space models and image space data are obtainable. Radiometric concerns take a more important role than previously. The data evaluation of the gray value of the digital image is no longer based on the digital image correlation. As an alternative, the gray values of an image are projected directly onto the models in the object space, this being a new principle. However, these numerical procedures in digital photogrammetry need to be stabilized by adjustment methods. Thus, the original concept of digital photogrammetry can be pragmatic to images from any sensor.

Considerable advances in digital photogrammetry have been made in recent years due to the availability of new hardware and software, such as image processing workstations and increased storage capacity [16, 17].

5. Device and acquisition characteristics

The main camera and photography parameters are focal length, focal point, bias, distortion, and pixel error; they will allow more accurate calibration [18] and are shown in **Figure 3**.

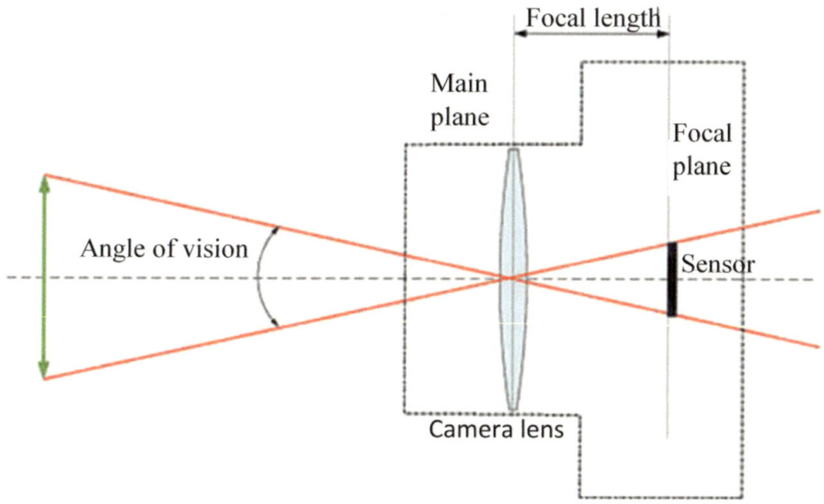

Figure 3.
Scheme of operation of a camera objective.

5.1 Camera objective

Included in the optical part of the camera, it is in charge in projecting the image that crosses it on the same plane and in outstanding conditions of sharpness. Therefore, it is a matter of focusing on the objects that are at equal distance on the focal plane. From certain distance, all the objects will be projected on the same plane. The light points are transmitted to an element that composes the scenario. As a result of diffraction, this is shown as a circular point with a halo around it and concentric rings, named Airy discs. Suppressing them is unfeasible because it is a physical light effect. Even so, it would be desirable for such rings to be as diffuse and thin as possible [17, 19].

Its resolving capacity depends on two parameters: aberrations and diffraction. One of the main functions of the objective is to suppress aberrations. When the diaphragm is closed, the aberrations are placated, and the only limiting factor is diffraction. When the diaphragm is opened, diffraction diminishes its significance in the wake of aberrations, which add up to force [20].

5.2 Focal length

This parameter is measured from the optical center of the lens to the focal plane, when the camera focused toward the infinity [5, 21]. Normal lenses are those which have a distance close to the diagonal of the cliché. The representation of the focal length is shown in **Figure 4**.

5.3 Relative aperture

Relative aperture (Ab) is the connection of the lens diameter (D) and its focal length (f) (Eq. (17)).

It is shown by the denominator, known as brightness or "f-number." In a different way, the aperture is the span through which light enters to be captured by the sensor. The more spacious the opening will be, the more light will enter the sensor as the number becomes smaller [4, 7]:

$$Ab = \frac{D}{f} \tag{17}$$

5.4 Field angle

This is the viewing angle of the camera and is closely related to the focal length and dimension of the sensor [8, 22]. A schematic representation is proposed in **Figure 5**.

5.5 Shutter

It is a mechanism that keeps the light passing through the lens into the closed camera. At certain intervals of time, it has the ability to open, allowing the passage of light so that the film can be impressed. The opening time can be set [21].

5.6 Focus depth

It is related to the permissiveness that occurred between obtaining a sharp image with a suitable impression and another less adequate exposure, although also producing a sharp image. Depth of focus is altered by lens magnification and numerical

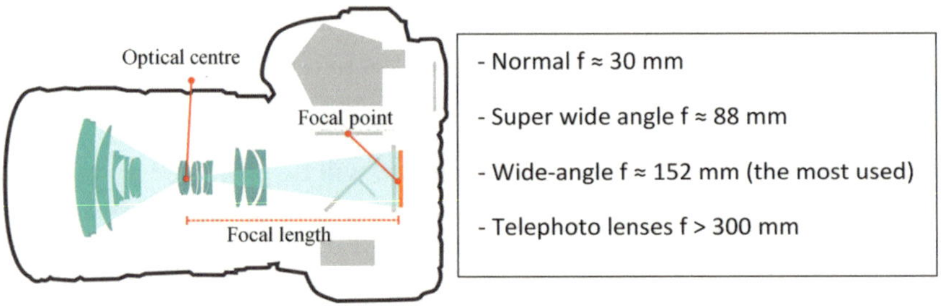

Figure 4.
Representation and focal length types on a camera.

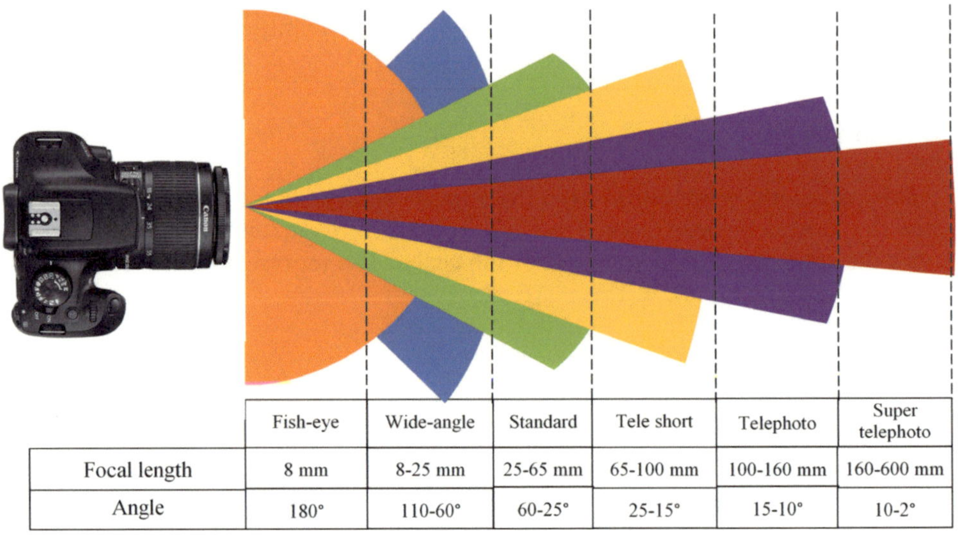

	Fish-eye	Wide-angle	Standard	Tele short	Telephoto	Super telephoto
Focal length	8 mm	8-25 mm	25-65 mm	65-100 mm	100-160 mm	160-600 mm
Angle	180°	110-60°	60-25°	25-15°	15-10°	10-2°

Figure 5.
Focal distances and corresponding angles.

aperture, and under some pretexts, large aperture systems have more pronounced depths of focus than low aperture systems, even if the depth of field is small [19].

5.7 Depth of field

Depth of field is the area of sharp reproduction seen in the photograph. In this one, there are some objects observed which are located at a certain distance, as well as others more distant or adjacent to them [20].

5.8 Sensor

Its function is to modify the light received in order to obtain a digital systematization. The sensor is called pixel in its minimum element. A digital image consists of a set of pixels. The technology based on complementary metal oxide semiconductor (CMOS) sensors is the most applied. The sensors consist of a semiconductor and sensitive material in the visible spectrum, between 300 and 1000 nm [10]. Charge-coupled device (CCD) sensors are becoming obsolete due to the cost and speed of processing images.

The comparison reading of the information in the CMOS sensors has the advantage of obtaining enough captures, obtaining readings using less time and with greater flexibility. Using a high dynamic range of work, high contrasts and a correct display of objects are achieved. In terms of quality, the physical size of the sensor is more significant than the number of cells or resolution. A large unit may allow higher-quality photographs to be taken than another sensor with a higher resolution but with a smaller surface [23].

As far as color is concerned, it must be seen that color is just a human visual perception. In order to be able to glimpse the color of an object, it is necessary to have a light source and something that reflects this light. A color is represented in digital format by applying a system of representation. The most commonly used is the RGB system. To represent a color, the exact percentages of primary red, primary green, and primary blue (RGB, red, green, blue) must be available. By this way, the color is displayed through the implementation of three numbers [24].

5.9 Diaphragm

The function of this element is to enlarge or decrease the percentage of light circulating through the target. The diaphragm aperture is related to the percentage of aperture it has. It is counted in f-numbers. The step is the shift from one value to the next. The ratio of luminosity, according to the scale of the f, does it in a factor of 2 [5] (**Figure 6**).

5.10 Other aspects to be taken into account

5.10.1 Focus

The first step in taking a picture is focusing. The most commonly used types of automatic focusing are [25]:

- Phase detection autofocus (PDAF). Its management is done by applying photodiodes through the sensor. The focusing element is moved in the lens to focus the image. It is a slow and inaccurate system due to the use of photodiodes.

- Dual pixel. This method uses more focus points along the sensor than the PDAF. This system uses two photodiodes at each pixel to compare minimal dissimilarities. This is the most effective focusing technology.

- Contrast detection. It is the oldest of the three systems exposed. Its operation theoretically bases that the contrast of an image is greater, and its edges are appreciated in a clearer way, when it is focused correctly. The disadvantage is its slowness.

5.10.2 Perspective

A photograph is a perspective image of an object. If straight lines are drawn from all points of an object to a fixed point (called point of view or center of projection) and lines are considered that cross an intermediate surface (called projection surface), the image is drawn on this surface and is known as perspective [1, 26].

The camera is responsible for executing and materializing perspectives of objects. The projection surface is the flat extension of the image sensor or the capture surface. Focal distance is the orthogonal distance separating the viewpoint from the projection surface. Knowing the distance between the point of view and the plane that contains the points of the object, the focal distance with which the photograph was taken and the inclination of the plane in which the points of the object to be measured are located with respect to the projection plane, the reliable coordinates of the points can be disintegrated, using basic trigonometry (**Figure** 7).

The orthogonal and the geometric perspectives are the most widely used in photogrammetry. Using a conventional camera (reel or digital), a geometric perspective will be plotted. From a photograph in which the points of the object to be measured are in a plane parallel to the projection plane or the one on which the photographic film is spread, the real position of the points in space is obtained by using Eqs. (18)–(19):

$$\frac{-x}{X} = \frac{X}{Z} \tag{18}$$

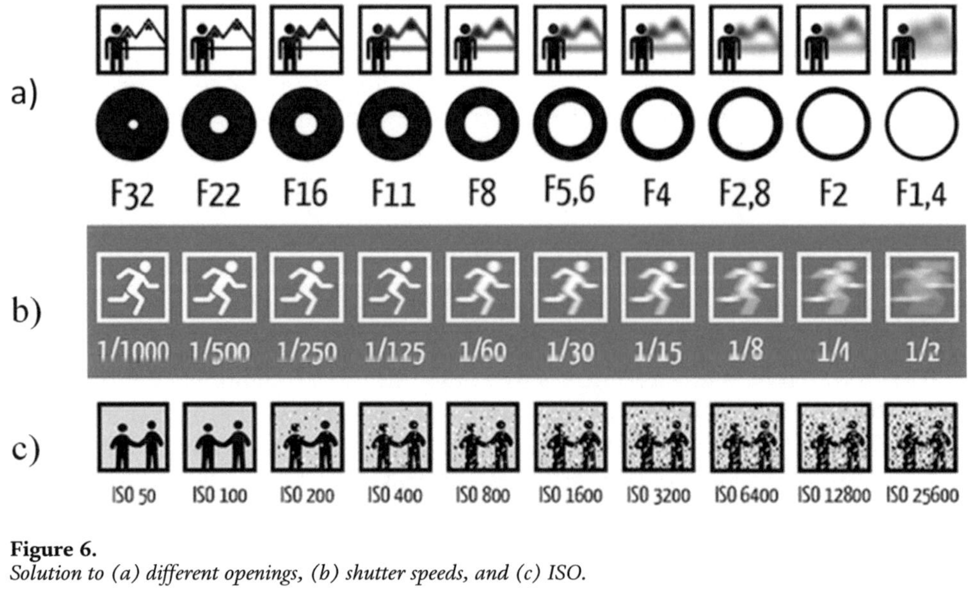

Figure 6.
Solution to (a) different openings, (b) shutter speeds, and (c) ISO.

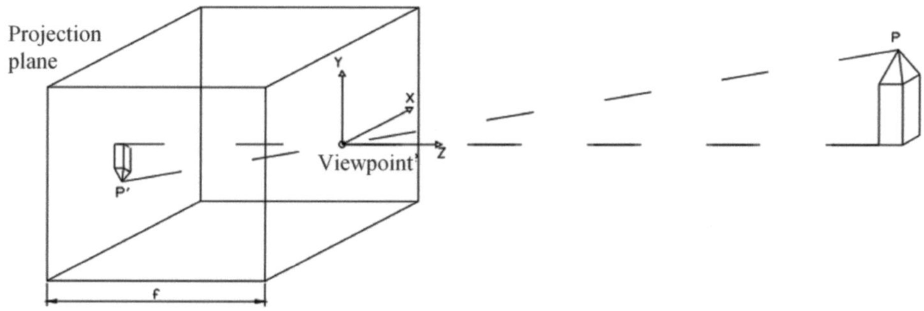

Figure 7.
Diagram of the projection of a camera.

$$\frac{-y}{f} = \frac{Y}{Z} \tag{19}$$

where *f* is the focal length, (*X, Y, Z*) are the actual coordinates of the point, and *P* (*x,y*) are the coordinates on the projection plane of the image or photograph.

It would be in front of more complex expressions if the planes that contain the points are not parallel to the one of projection, being indispensable to know the inclination of the plane having as reference the plane of projection. In practice, in order to avoid complications in the calculation of coordinates, photographs are usually taken in a way that the planes are parallel.

5.10.3 Exposure

It is based on the capture of a scene by means of a sensitive material. In analog photography, this corresponds to the film and in digital photography, the sensor. Exposure is based on three variables to control the entry of light into the focal plane (sensor) and achieve an adequate exposure [9]:

1. ISO Sensitivity: it indicates the amount of light required to take a picture. The higher the light, the lower the ISO.

2. Diaphragm opening: it inspects the light reaching the focal plane, along with the shutter speed, and regulates the depth of field of the photograph.

3. Shutter speed: shutter opening time allows light to reach the sensor. The higher the shutter speed, the lower the percentage of light reaching the sensor.

When a sensor has the ability to capture as many tones (dynamic range) and information (light) as its ability allows, the picture is perfectly exposed.

5.10.4 Dynamic range

It measures the amount of light and dark tones that a camera has the ability to capture in the same picture. It shows the amount of tonal nuances that a camera is capable of capturing, measurable by contrast and sharpness.

Contrast and sharpness are based on the differentiation of tonality with which a pair of white and black lines are obtained, captured, or reproduced. It is measurable of the degree of detail, being 100% when both lines can be perfectly differentiated as pure whites and blacks. Resolution and contrast are closely related concepts.

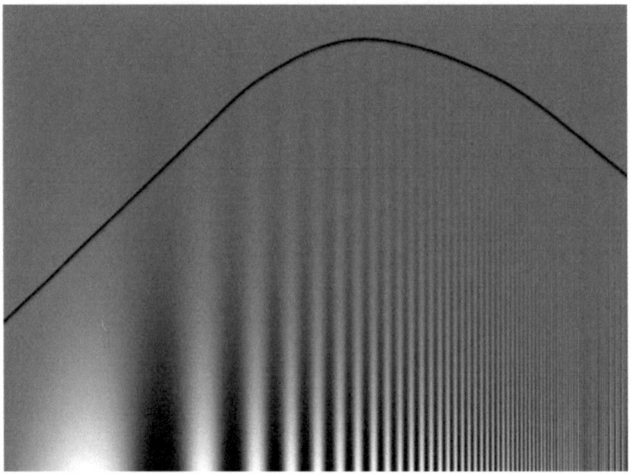

Figure 8.
Contrast sensitivity change as a function of the spatial frequency of the target.

If the contrast falls below 5%, it is difficult to observe any detail, which is shown more clearly and distinctly the higher it is. Frequency and modulation are shown in the way they are altered when light passes through the different optical components of the lens of the photographed image, thanks to contrast transfer functions. As the viewer moves away, a substantial loss of contrast begins to be noticed [12].

By performing a contrast correction, different filters are applied to the central zones instead of the peripheral zones. An example of contrast and resolution is shown in **Figure 8**.

5.10.5 Aberrations

One of the most outstanding components of a camera is the photographic lens, which produces a series of aberrations that distort the images of the photographs, making difficult to visualize the correct dimensions of the object [27, 28]. There are different types of aberrations, being the most common in photographic lenses:

1. Point aberrations: housed in the position arranged by the paraxial optics. It is a "stain" instead of a point. There are also chromatic aberration, spherical aberration, astigmatism, and coma.

2. Shape aberrations: the point is shown as a point but with a different position to the one arranged by means of paraxial approximation. This is a systematic error and can be of two types: field curvature and distortions.

    - Field curvature: defect when creating the image, being curved instead of flat. It is difficult to correct the aberration, but it can be mitigated in a low percentage.

    - Distortion: only affects the shape of the image. It occurs due to the difference in the scale of reproduction of the image off-axis. If an object with straight lines is photographed, such as a square, the center lines will appear straight, and the edge lines will curve inward or outward producing the so-called barrel or cushion distortions. This aberration is not corrected by closing the diaphragm. This error affects the tone of the image and needs to be corrected.

5.10.6 Environmental conditions

Stability of environmental conditions must be achieved:

1. Temperature: the ideal temperature for taking a photograph should be between approximately 18 and 26° in order to avoid dilatation of the lens.

2. Wind: calm wind, to avoid hindrances when taking the photo.

3. Illumination: sufficient light bulb. In most cases, natural light is not sufficient, and it is necessary to use spotlights or other artificial elements.

Other significant parameters, such as the texture of the element, significantly help the quality of the 3D reconstruction, and optimal results are obtained with the highest level of ambient light (exposure 1/60, f/2.8, and ISO sensitivity 100). The surface of an element should be opaque, with Lambertian reflection and surface homogeneity. A single point on the surface of the object must be visible from at least two or more sensors [26, 29].

5.10.7 Image quality

Image quality is a prerequisite for working with it properly. There are two main characteristics that define it:

1. Resolution in amplitude (bit depth): number of bits per point of an image

2. Spatial resolution: the number of pixels per unit area

Image processing is the transformation of an input image into an output image. It is carried out to facilitate the analysis of the image and to obtain a greater reliability of this [30]. Among the transformations, those that eliminate noise or variation in the intensity of the pixels stand out. There are two types of operations: individual operations (rectification or binarization) and neighborhood operations (filtering).

5.10.8 Histogram

This is a visual tool very useful for the study of digital images. With the naked eye, it is possible to study the contrast or the distribution of intensities, because it follows the following discrete function of Eq. (20):

$$f(x_i) = n_i \tag{20}$$

where x is the level of gray or color and n is the number of pixels in the image with this value. The histogram is normalized in values ranging from 0 to 1. In **Figure 9** it is possible to see their different zones [18, 31].

The most common errors in the image, which prevent good image quality, can be identified in the histogram and are muted tones, black areas, overexposure or burned areas, and backlight. In order to know that a good image is acquired, the best thing is to have a histogram that has the shape of a Gauss bell, that is to say, that has the most information in the central part and less in the extremes. Another important point is that the histogram must embrace and reach both ends, so as to ensure that there are blacks and whites in the photograph.

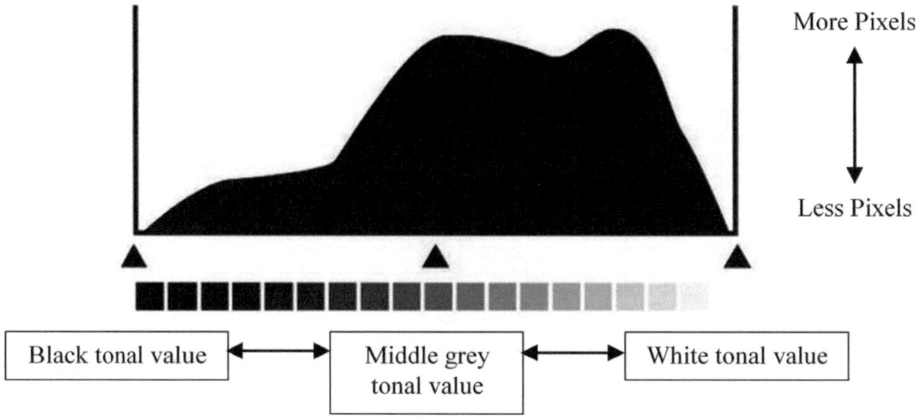

Figure 9.
Histogram areas.

5.10.9 Binarization

The representation of an image with two values is obtained. The dimensions of the image are still preserved. The decision threshold must be chosen correctly and used in a step filter with an algorithm similar to Eq. (21):

$$g(x,y) = \begin{cases} 0 & f(x,y) > k \\ 1 & f(x,y) \leq k \end{cases} \tag{21}$$

where 0/1 represents the black/white values and f is the value of the gray tone of the coordinates (x,y) [32]. **Figure 10** shows a grayscale image versus a binary.

To obtain an image with sufficient quality, the binarization must correspond with white pixels to the objects of interest, being the blacks of the environment. If the object of interest turns out to be darker than the environment, a reversal is applied after the binarization. The most important point in the process is the calculation of the threshold. There are different methods for this: histogram, clustering, entropy, similarity, spatial, global, and local.

The setting of the threshold value is latent, due to its difficulty, in all methods. The techniques are supported by statistics applied to the histogram. They are as follows: carry error method, Otsu method, and Saulova's pixel deviation method.

5.10.10 Spatial filtering

It is based on a convolution operation between the two-dimensional functions image, f, and a nucleus, called h, in digital images. This operation aims to transform

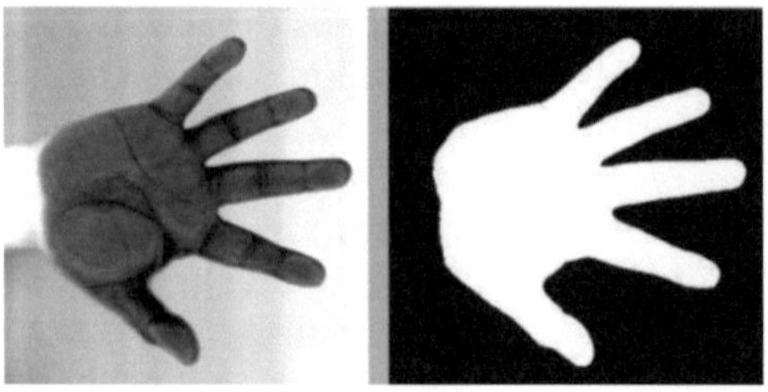

Figure 10.
Grayscale (left) and binary (right).

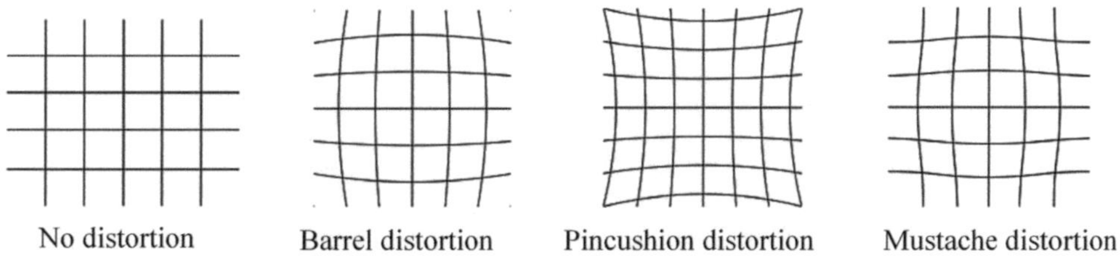

Figure 11.
Types of lens distortion.

the value of a pixel p into the position (x,y), always taking into account the values of the adjacent pixels. For this operation, a weighted sum of the values of the neighboring points of this point p is required. A mask (h), behaving like a filter, is in charge of exposing the values of the weighting. The size of the mask varies according to the pixels used.

5.10.11 Geometrical transformations

These operations modify the spatial coordinates of the image. There are several operations that are easy to understand and apply, such as interpolation, rotation, rectification, and distortion correction.

5.10.12 Lens distortion

Due to the geometry of the lens, it reproduces a square object with variations in its parallel lines. There are three types of distortion: barrel, pincushion, and mustache (combination of the first two) (**Figure 11**) [25, 33]. This error is negligible in a photograph of a natural scene, but to take engineering measurements and obtain a virtual object, it is necessary to compensate for the distortion. There is a mathematical model for the treatment of distortion.

The barrel distortion is centered and symmetrical. Therefore, to correct the distortion of a certain point, a radial transformation is performed, expressed mathematically in Eq. (22):

$$\begin{pmatrix} \hat{x} - x_d \\ \hat{y} - y_d \end{pmatrix} = L\sqrt{(x-x_d)^2 + (y-y_d)^2}\begin{pmatrix} x - x_d \\ y - y_d \end{pmatrix} \tag{22}$$

where $(\hat{x}, \hat{y})$ represents the result of the distortion correction at point (x, y), (x_d, y_d) represents the center of the distortion which is usually a point near the center of the image, and finally the radial function $L(r)$ determines the magnitude of the distortion correction as a function of the distance from the point to the center of distortion [34].

The radial function $L(r)$ is performed by applying two strategies. The first one gives rise to the so-called polynomial models (Eq. (23)):

$$L(r) = 1 + k_1 r^2 + k_2 r^4 + \ldots + k_n r^{2n} \tag{23}$$

The second one is based on an approach (Eq. (24)):

$$L(r) = \frac{1}{1 + k_1 r^2 + k_2 r^4 + \ldots + k_n r^{2n}} \tag{24}$$

Figure 12.
Visual example of photo rectification.

The values k_1–k_n are called distortion model parameters. These values, together with the distortion center coordinates (x_d, y_d), completely represent the distortion model. The distortion of the lens is represented by the k_i coefficients. They are obtained from a known calibration image.

5.10.13 Rectification (perspective distortion)

Image correction is necessary because either it is difficult to keep the optical axis vertical at all points of the shot or the axis is tilted toward the vertical. Vertical images are obtained free of displacement because of the inclination of the shot but still have inclinations, product of the depth of the workpiece. Displacements can be suppressed by applying differential grinding or orthorectification process. In the original digital image or a scan, the technique is applied pixel by pixel. In a scanned image, the initial data are the coordinates of the control points. The procedure is divided into two steps:

1. Determination of the mathematical transformation related to real coordinates and those belonging to the image

2. Achievement of new image, being aligned to the reference system

After this process, it is necessary to know that all the pixels of the resulting orthophotography have their level of gray, performing a digital resampling [17, 34]. **Figure 12** shows an unrectified (left) and rectified (right) photograph.

Several resamples are made on the initial image. Three resampling methods are regularly used: bilinear interpolation, nearest neighbor, and bicubic convolution. The transformations to be applied to the images are [19] Helmert transformation; affine transformation; polynomial transformation; and two-dimensional projective transformation.

6. Obtaining a 3D model from 2D photographs

To obtain a 3D model of an object from a 2D one, photographs must be taken from different views, with adequate quality. From these photographs, the reconstruction process begins.

3D reconstruction is the process by which real objects are reproduced on a computer. Nowadays there are several reconstruction techniques and 3D mesh methods, having a function to obtain an algorithm that is able to make the connection of the set of representative points of the object in form of surface elements.

The efficiency with which the techniques are used will be linked to the final quality of the reconstruction.

The stereoscopic scene analysis system presented by Koch uses image matching, object segmentation, interpolation, and triangulation techniques to obtain the 3D point density map. The system is divided into three modules: sensor processing, image pair processing, and model-based sequence processing.

Pollefeys features a 3D reconstruction process based on well-defined stages. The input is an image sequence, and the output of the process is a 3D surface model. The stages are the following: image ratio, structure and motion recovery, dense matching, and model construction.

Another proposal is expressed by Remondino. He presents a 3D reconstruction system following these steps: image sequence acquisition and analysis, image calibration and orientation, matching process and the generation of points, and 3D modeling [18].

6.1 From a photograph

It is used in revolutionary pieces. With only one photograph, it is possible to obtain the axis and dimensions. In 1978 Barrow and Tenenbaum demonstrated that the orientation of the surface along the silhouette can be calculated directly from the image data, resulting in the first study of silhouettes in individual views. Koenderink showed that the sign of the silhouette's curvature is equivalent to that of the Gaussian curvature. Thus, concavities, convexities, and inflections of the silhouette indicate hyperbolic, convex, and parabolic surface points, respectively. Finally, Cipolla and Blake exposed that the curvature of the silhouette has the corresponding sign as the normal curvature along the contour generator in the perspective projection. A similar result was derived for the orthographic projection by Brady [35].

First, the silhouette ρ of a surface of revolution (SOR) is extracted from the image with a Canny edge detector, and the harmonic homology W that maps each side of ρ to its symmetrical complement is predictable by minimizing the geometric detachments among the original silhouette ρ and its transformation version $\rho' = W\rho$. The image is rectified, and the axis of the figure is rotated and put in orthogonal projection (**Figure 13**).

The apparent contour is first manually segmented from the rectified silhouette. This can usually be done easily by removing the upper and lower elliptical parts of the silhouette. The points are then sampled from the apparent contour, and the tangent vector (i.e., $\dot{x}(s)$ *and* $\dot{y}(s)$) at each sample point is calculated by fitting a polynomial to the neighboring points.

For $\Psi \neq 0$, $R_x(\Psi)$ first transforms the display vector $p(s)$ and the associated surface normal $n(s)$ at each sample point: the transformed display vector is normalized so that its third coefficient becomes one, and the following Eqs. (25)–(26) can be used to recover the depth of the sample point:

$$n(s) = \frac{1}{\alpha_n(s)} \begin{bmatrix} -\dot{y}(s) \\ \dot{x}(s) \\ x(s)\dot{y}(s) - \dot{x}(s)y(s) \end{bmatrix} \tag{25}$$

where

$$\alpha_n(s) = \left| p(s)x\frac{dp(s)}{ds} \right| \tag{26}$$

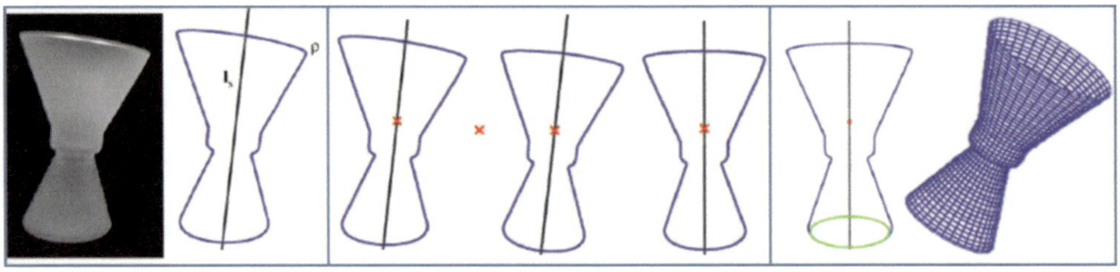

Figure 13.
Harmonic homology of the figure and its transformation to orthogonal projection [35].

6.2 From two photographs

This section is based on an investigation using a practical heuristic method, for the reconstruction of structured scenes from two uncalibrated images. The method is based on an initial estimation of the main homographies of the initial 2D point coincidences, which may contain some outliers, and the homographies are recursively refined by incorporating the point and line support coincidences on the main spatial surfaces. The epipolar geometry is then recovered directly from the refined homogenies, and the chambers are calibrated from three orthogonal vanishing points, and the infinite homography is recovered.

First, a simple homography-guided method is proposed to fit and match the line segments between two views, using Canny edge detector and regression algorithms. Second, the cameras are automatically calibrated with the four intrinsic parameters that vary between the two views. A RANSAC mechanism is adopted to detect the main flat surfaces of the object from 2D images. The advantages of the method are that it can build more realistic models with minimal human interactions and it also allows more visible surfaces to be reconstructed on the detected planes than traditional methods that can only reconstruct overlapping parts (**Figure 14**).

6.3 Though more than two photographs

6.3.1 Reconstruction of geological objects

This is one of the fields where photogrammetry is most applied nowadays. In this specific point, the reconstruction is carried out applying Delaunay's triangulation and the tetrahedron. Many data models based on tetrahedron mesh have been developed to represent the complex objects in 3D GIS.

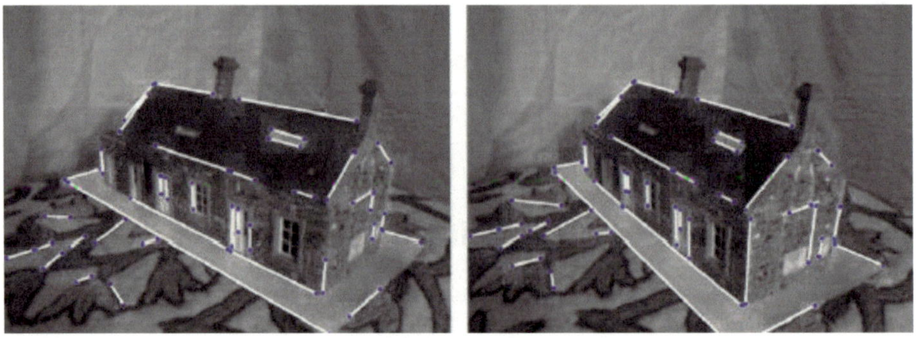

Figure 14.
The matching results of the line segments in four main planes [36].

The tetrahedron grid can only be used to represent the geometrical structure of geological objects. The natural characteristics of geological objects are reflected in their different attributes, such as different rock formations, different contents of mineral bodies, etc. It is defined that the attribute value of the internal point can be linearly interpolated from the attribute values in four vertices in a tetrahedron. But the attributes could change suddenly between different formations and different mineral bodies. To cope with sudden changes, interpolation of the tetrahedron is needed that can only be applied to six sides of a tetrahedron. Those interpolated points are only used as time data for the following processing [37].

6.3.2 Reconstruction of objects with high surface and texture resolution

This section presents a robust and precise system for the 3D reconstruction of real objects with shapes and textures in high resolution. The reconstruction method is passive, and the only information required is 2D images obtained with a camera calibrated from different viewing angles as the object rotates on a rotating plate. The triangle surface model is obtained through a scheme that combines the octree construction and the walking cube algorithm. A texture mapping strategy based on surface particles is developed to adequately address photographic-related problems such as inhomogeneous lighting, lights, and occlusion [38]. To conclude, the results of the reconstruction are included to demonstrate the quality obtained (**Figure 15**).

The scheme combining octree construction and isolevel extraction through marching cubes is presented for the problem concerning the shape of the silhouette. The use of octree representation allows to reach very high resolutions, while the method of fast walking cubes is adapted through a properly defined isolevel function to work with binary silhouettes, resulting in a mesh of triangles with vertices precisely located in the visual object.

Calibration is performed on the camera and rotary table. One of the problems found is the discontinuity of the texture due to the nonhomogeneous lighting in different parts of the element due to shadows.

Next, the octree is represented. An octree is a hierarchical tree structure that can be used to represent volumetric data in terms of cubes of different sizes. Each octree node corresponds to a cube in the octree space that is entirely within the object. This opens up different possibilities: voxels, particles, triangles, and more complicated

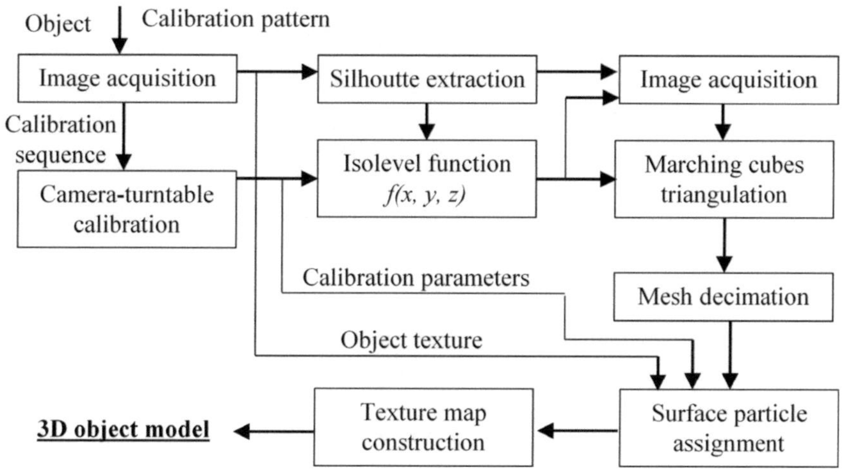

Figure 15.
Flowchart to the reconstruction of objects.

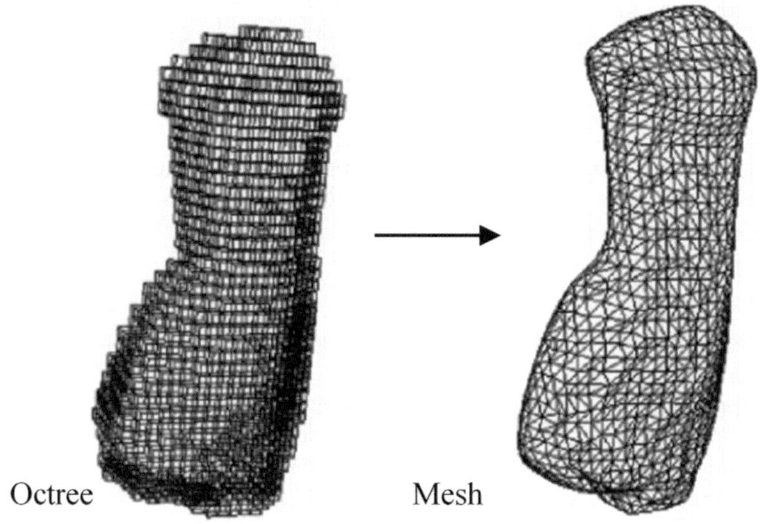

Figure 16.
From cube to triangulation, adapted from [38].

parametric primitives, such as splines or NURBS. Voxels are used to represent volumes but can also be used to represent surfaces. A related primitive is a particle that is defined by its color, orientation, and position. By marching the cube triangulation of the octree, the white and black points denote the corners of the cube that are inside and outside, respectively, while the gray points are the points of the triangle's vertex on the surface (**Figure 16**).

The application of the isolevel function calculated by means of the dichotomous subdivision procedure allows for the construction of a faithful model of the object. The triangular vertices that make up the object's mesh are placed precisely on the surface of the digitized model even at low resolutions. This creates an efficient compromise between resolution and geometric accuracy. The octree construction followed by the walking cube algorithm generates a triangular mesh consisting of an excessive number of triangles, which must be simplified.

6.3.3 Object reconstruction

The reconstruction of objects is mainly based on the archeological field. The process to obtain the 3D model will be governed by **Figure 17**.

First of all, corresponding or common characteristics must be found among the images of the object. The process occurs in two phases:

1. The reconstruction algorithm generates a reconstruction in which dimensions are not correctly defined. A self-calibration algorithm performs a reconstruction equivalent to the original one, formed by a set of 3D points.

2. All the pixels of an image are made to coincide with those of the neighboring images so that the system can reconstruct these points.

The system selects two images to set up an initial projective reconstruction frame and then reconstructs the matching feature points through triangulation.

Then a dense surface estimation is performed. To obtain a more detailed model of the observed surface, a dense matching technique is used. The 3D surface is

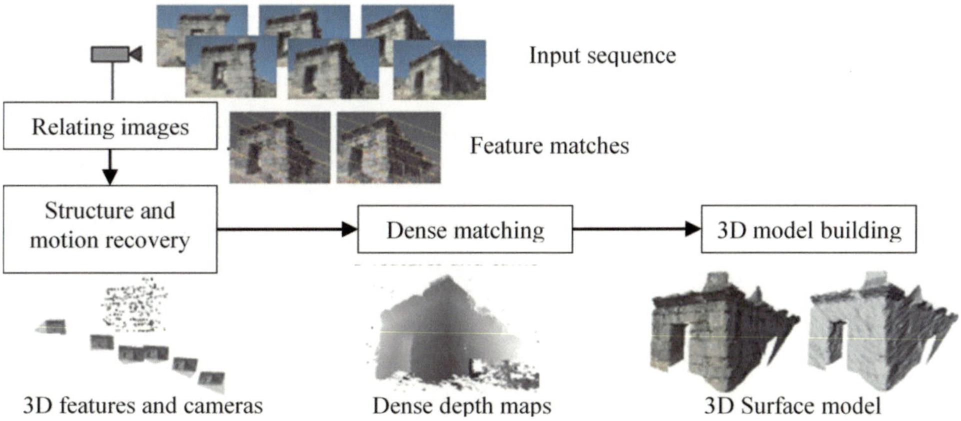

Figure 17.
Steps to obtain the 3D model, adapted from [39].

approached with a triangular grid, to reduce geometric complexity and adapt the model to the requirements of the computer graphic display system. Then construct a corresponding 3D mesh by placing the triangle vertices in 3D space according to the values found in the corresponding depth map. To reconstruct more complex shapes, the system must combine multiple depth maps. Finally, it is provided with texture.

6.3.4 3D reconstruction of the human body

It is used for medical purposes in many cases, as a base for implants, splints, etc. The process consists of the following parts: acquisition and analysis of the image sequence; calibration and orientation of the images; matching process on the surface of the human body; and generation and modeling of the point cloud. Once the necessary images have been obtained from different points of view, the calibration and orientation of the images are carried out.

The choice of the camera model is often related to the final application and the required accuracy. The correct calibration of the sensor used is one of the main objectives. Another important point is image matching [40].

To evaluate the quality of the matching results, different indicators are used: an ex post standard deviation of the least squares adjustment, the standard deviation of the change in the x-y directions, and the shift from the initial position in the x-y directions. The performance of the process, in the case of uncalibrated images, can only be improved with a local contrast enhancement of the images.

Finally, 3D reconstruction and modeling of the human body shape is performed. The 3D coordinates of each matching triplet are calculated through a forward intersection. Using collinearity and the results of the orientation process, the 3D paired points are determined with a solution of least squares. For each triplet of images, a point cloud is calculated, and then all the points are joined together to create a unique point cloud. A spatial filter is applied to reduce noise and obtain a more uniform point cloud density. **Figure 18** shows the results before and after filtering (approximately 20,000 points, left); a view of the recovered point cloud with pixel intensity (center); and a 3D human model (right).

The system is composed of two main modules. The first one is in charge of image processing, to determine the depth map in a pair of views, where each pair of successive views follows a sequence of phases: detection of points of interest, correspondence of points, and reconstruction of these. In this last phase, the

Figure 18.
3D reconstruction of a human body, adapted from [40].

parameters that describe the movement (rotation matrix R and translation vector T) between the two views are determined. This sequence of steps is repeated for all successive pairs of views of the set.

The second module is responsible for creating the 3D model, for which it must determine the total 3D points map generated. In each iteration of the previous module, the 3D mesh is generated by applying Delaunay's triangulation method. The results obtained from the process are modeled in a virtual environment to obtain a more realistic visualization of the object [16].

The number of detected minutiae is related to the number of reconstructed 3D points and the quality of that reconstruction (higher number of details). Therefore, the higher the number of points on the map, the more detailed areas are obtained. In some cases this does not apply, due to the geometry of the object, for example, in a cube, more points can result in a distorted object.

7. Conclusion

The technological development of 3D photogrammetry makes it a real option in the various applications of 3D scanners. Among the different benefits it brings are faster raw data acquisition, simplicity, portability, and more economical equipment. Different studies have verified the accuracy and repeatability of 3D photogrammetry. These investigations have compared the digital models of objects obtained from 2D digital photographs with those generated by a 3D surface scanner. In general, the meshes obtained with photogrammetric techniques and with scanners show a low degree of deviation from each other. The surface settings of photogrammetric models are usually a little better. For these reasons, photogrammetry is a technology with an infinite number of engineering applications.

In this chapter the basic fundamentals, the characteristics of the acquisition, and the aspects to be taken into account to obtain a good virtual model from photogrammetry have been explained.

Acknowledgements

The authors would like to thank the call for Innovation and Teaching Improvement Projects of the University of Cadiz and AIRBUS-UCA Innovation Unit (UIC) for the Development of Advanced Manufacturing Technologies in the Aeronautical Industry.

Conflict of interest

The authors declare no conflict of interest.

Author details

Ana Pilar Valerga Puerta[1*], Rocio Aletheia Jimenez-Rodriguez[2], Sergio Fernandez-Vidal[2] and Severo Raul Fernandez-Vidal[1]

1 Department of Mechanical Engineering and Industrial Design, School of Engineering, University of Cadiz, Cadiz, Spain

2 Reanimalia. Rehabilitacion and Ortopedia Veterinaria, Cadiz, Spain

*Address all correspondence to: anapilar.valerga@uca.es

References

[1] Rolin R, Antaluca E, Batoz JL, Lamarque F, Lejeune M. From point cloud data to structural analysis through a geometrical hBIM-oriented model. Journal of Cultural Heritage. 2019;**12**: 1-26. DOI: 10.1145/3242901

[2] Valerga AP, Batista M, Bienvenido R, Fernández-Vidal SR, Wendt C, Marcos M. Reverse engineering based methodology for modelling cutting tools. Procedia Engineering. 2015;**132**: 1144-1151. DOI: 10.1016/j.proeng.2015.12.607

[3] Rabbani T, Dijkman S, van den Heuvel F, Vosselman G. An integrated approach for modelling and global registration of point clouds. ISPRS Journal of Photogrammetry and Remote Sensing. 2007;**61**:355-370. DOI: 10.1016/j.isprsjprs.2006.09.006

[4] Schenk T. Introduction to Photogrammetry. Department of Civil and Environmental Engineering and Geodetic Science. Athens, USA: The Ohio State University; 2005. pp. 79-95. Available from: http://gscphoto.ceegs.ohio-state.edu/courses/GeodSci410/docs/GS410_02.pdf

[5] Derenyi EE. Photogrammetry: The Concepts. Canada: Department of Geodesy and Geomatics Engineering University of New Brunswick; 1996. DOI: 10.1017/9781108665537.002

[6] Ackermann F. Digital image correlation: Performance and potential application in photogrammetry. The Photogrammetric Record. 1984;**11**: 429-439. DOI: 10.1111/j.1477-9730.1984.tb00505.x

[7] Sansoni G, Trebeschi M, Docchio F. State-of-the-art and applications of 3D imaging sensors in industry, cultural heritage, medicine, and criminal investigation. Sensors. 2009;**9**:568-601. DOI: 10.3390/s90100568

[8] Katz D, Friess M. Technical note: 3D from standard digital photography of human crania - A preliminary assessment. American Journal of Physical Anthropology. 2014;**154**: 152-158. DOI: 10.1002/ajpa.22468

[9] Martorelli M, Lepore A, Lanzotti A. Quality analysis of 3D reconstruction in underwater photogrammetry by bootstrapping design of experiments. International Journal of Mechanical Sciences. 2016;**10**:39-45

[10] Guerra MG, Lavecchia F, Maggipinto G, Galantucci LM, Longo GA. Measuring techniques suitable for verification and repairing of industrial components: A comparison among optical systems. CIRP Journal of Manufacturing Science and Technology. 2019;**27**:114-123. DOI: 10.1016/j.cirpj.2019.09.003

[11] Givi M, Cournoyer L, Reain G, Eves BJ. Performance evaluation of a portable 3D imaging system. Precision Engineering. 2019;**59**:156-165. DOI: 10.1016/j.precisioneng.2019.06.002

[12] Aguilar R, Noel MF, Ramos LF. Integration of reverse engineering and non-linear numerical analysis for the seismic assessment of historical adobe buildings. Automation in Construction. 2019;**98**:1-15. DOI: 10.1016/j.autcon.2018.11.010

[13] Murphy M, McGovern E, Pavia S. Historic building information modelling - Adding intelligence to laser and image based surveys of European classical architecture. ISPRS Journal of Photogrammetry and Remote Sensing. 2013;**76**:89-102. DOI: 10.1016/j.isprsjprs.2012.11.006

[14] Dai F, Lu M. Assessing the accuracy of applying photogrammetry to take geometric measurements on building products. Journal of Construction

Engineering and Management. 2010; **136**:242-250. DOI: 10.1061/(ASCE) CO.1943-7862.0000114

[15] Wrobel BP. The evolution of digital photogrammetry from analytical photogrammetry. The Photogrammetric Record. 1991;**13**:765-776. DOI: 10.1111/ j.1477-9730.1991.tb00738.x

[16] Styliadis AD, Sechidis LA. Photography-based façade recovery & 3-d modeling: A CAD application in cultural heritage. Journal of Cultural Heritage. 2011;**12**:243-252. DOI: 10.1016/j.culher.2010.12.008

[17] Murtiyoso A, Grussenmeyer P, Börlin N. Reprocessing close range terrestrial and uav photogrammetric projects with the dbat toolbox for independent verification and quality control. International Archives of the Photogrammetry, Remote Sensing and Spatial Information Sciences - ISPRS Archives. 2017;**42**:171-177. DOI: 10.5194/isprs-archives-XLII-2-W8-171-2017

[18] Remondino F, Fraser C. Digital camera calibration methods: Considerations and comparisons. The International Archives of the Photogrammetry, Remote Sensing and Spatial Information Sciences. 2006;**36**: 266-272

[19] Bister D, Mordarai F, Aveling RM. Comparison of 10 digital SLR cameras for orthodontic photography. Journal of Orthodontics. 2006;**33**:223-230. DOI: 10.1179/146531205225021687

[20] Bill Triggs AWF, McLauchlan PF, Hartley RI. In: Triggs B, editor. Bundle Adjustment—A Modern Synthesis Bill. Vision Algorithms '99. LNCS 1883; **2000**:298-372. DOI: 10.3760/cma.j. issn.2095-4352.2016.06.018

[21] Veal CJ, Holmes G, Nunez M, Hoegh-Guldberg O, Osborn J. A comparative study of methods for surface area and three dimensional shape measurement of coral skeletons. Limnology and Oceanography: Methods. 2010;**8**:241-253. DOI: 10.4319/ lom.2010.8.241

[22] Kaufman J, Clement M, Rennie AE. Reverse engineering using close range photogrammetry for additive manufactured reproduction of Egyptian artifacts and other Objets d'art (ESDA1014-20304). Journal of Computing and Information Science in Engineering. 2015;**15**:1-7. DOI: 10.1115/ 1.4028960

[23] Bernard A. Reverse engineering for rapid product development: A state of the art. Three-Dimensional Imaging, Optical Metrology, and Inspection. 1999; **3835**:50-63. DOI: 10.1117/12.370268

[24] Adeline KRM, Chen M, Briottet X, Pang SK, Paparoditis N. Shadow detection in very high spatial resolution aerial images: A comparative study. ISPRS Journal of Photogrammetry and Remote Sensing. 2013;**80**:21-38. DOI: 10.1016/j.isprsjprs.2013.02.003

[25] Yi XF, Long SC. Precision displacement measurement of single lens reflex digital camera. Applied Mechanics and Materials. 2011;**103**: 82-86. DOI: 10.4028/www.scientific. net/AMM.103.82

[26] Webster C, Westoby M, Rutter N, Jonas T. Three-dimensional thermal characterization of forest canopies using UAV photogrammetry. Remote Sensing of Environment. 2018;**209**:835-847. DOI: 10.1016/j.rse.2017.09.033

[27] Galantucci LM, Lavecchia F, Percoco G, Raspatelli S. New method to calibrate and validate a high-resolution 3D scanner, based on photogrammetry. Precision Engineering. 2014;**38**:279-291. DOI: 10.1016/j.precisioneng.2013.10.002

[28] Menna F, Nocerino E, Remondino F. Optical aberrations in

underwater photogrammetry with flat and hemispherical dome ports. In: Videometrics, Range Imaging, and Applications XIV. SPIE Optical Metrology. Munich, Germany: SPIE;. 2017;**1033205**:1-14. DOI: 10.1117/12.2270765

[29] Nevalainen O, Honkavaara E, Tuominen S, Viljanen N, Hakala T, Yu X, et al. Individual tree detection and classification with UAV-based photogrammetric point clouds and hyperspectral imaging. Remote Sensing. 2017;**9**:1-34. DOI: 10.3390/rs9030185

[30] Yoo Y, Lee S, Choe W, Kim C. CMOS image sensor noise reduction method for image signal processor in digital cameras and camera phones Youngjin. In: Proceedings of SPIE-IS&T Electronic Imaging Digital Photography III. 2007. pp. 1-10. DOI: 10.1117/12.702758

[31] Fliegel K, Havlin J. Imaging photometer with a non-professional digital camera. In: SPIE 7443, Applications of Digital Image Processing XXXII. 2009. pp. 1-8. DOI: 10.1117/12.825977

[32] Gomez-Gil P. Shape-based hand recognition approach using the morphological pattern spectrum. Journal of Electronic Imaging. 2009;**18**: 13012. DOI: 10.1117/1.3099712

[33] Ng R, Hanrahan PM. Digital correction of lens aberrations in light field photography. In: International Optical Design Conference. 2006. p. 6342. DOI: 10.1117/12.692290

[34] Jianping Z, John G. Image pipeline tuning for digital cameras. In: IEEE International Symposium on Consumer Electronics. Irving, TX; 2007. pp. 167-170

[35] Wong KYK, Mendonça PRS, Cipolla R. Reconstruction of surfaces of revolution from single uncalibrated views. Image and Vision Computing. 2004;**22**:829-836. DOI: 10.1016/j.imavis.2004.02.003

[36] Wang G, Tsui HT, Hu Z. Reconstruction of structured scenes from two uncalibrated images. Pattern Recognition Letters. 2005;**26**:207-220. DOI: 10.1016/j.patrec.2004.08.024

[37] Xue Y, Sun M, Ma A. On the reconstruction of three-dimensional complex geological objects using Delaunay triangulation. Future Generation Computer Systems. 2004; **20**:1227-1234. DOI: 10.1016/j.future.2003.11.012

[38] Yemez Y, Schmitt F. 3D reconstruction of real objects with high resolution shape and texture. Image and Vision Computing. 2004;**22**:1137-1153. DOI: 10.1016/j.imavis.2004.06.001

[39] Pollefeys M, Van Gool L, Vergauwen M, Cornelis K, Verbiest F, Tops J. 3D recording for archaeological fieldwork. IEEE Computer Graphics and Applications. 2003;May/June:20-27. DOI: 10.1109/MCG.2003.1198259

[40] Remondino F. 3-D reconstruction of static human body shape from image sequence. Computer Vision and Image Understanding. 2004;**93**:65-85. DOI: 10.1016/j.cviu.2003.08.006

Section 2

Product Development

Chapter 4

Virtual Prototyping Platform for Designing Mechanical and Mechatronic Systems

Cătălin Alexandru

Abstract

The chapter deals with the description of a virtual prototyping platform that facilitates the design process of the mechanical and mechatronic systems. The virtual prototyping stages are defined and then integrated in a block diagram, highlighting how the data are transferred between these stages in order to finally obtain a valid and optimal virtual model, close (as structure and functionality) to the real one. The whole process is guided by the basic principle for successful virtual prototyping: as complicated as necessary and as simple as possible. The real modeling case, the specific simplifying assumptions, and the validity (viability) fields of the simplifying assumptions are discussed with reference to the main components of a mechanical or mechatronic system (bodies, connections between bodies, actuating elements). The purpose is to manipulate the simplifying assumptions in a way that reduces the complexity of the virtual model, but without altering the accuracy of the results. The basic types of analysis/simulation are depicted by considering their particularities, highlighting their role in the process of designing mechanical/mechatronic systems, and then the optimization is conducted by the use of parametric design tools. Finally, a case study is developed following those mentioned above.

Keywords: virtual prototyping, mechanical and mechatronic systems, modeling, simulation, optimal design

1. Introduction

In the process of product development, as in many other fields, using the computer is no longer just a useful alternative to classical instruments, but it has become a real need. The computer is currently used from the concept elaboration stage until the manufacturing and implementation. Designers now have access to very sophisticated and high-performance working tools, based on software solutions dedicated to the various stages of product design and development. The traditional computer-aided design (CAD) and computer-aided manufacturing (CAM) approaches are now being addressed through computer-aided engineering (CAE) integrating platforms, which allow the evaluation and improvement of the product at the system level and not separately on its parts or subsystems, such an approach being reflected in increasingly efficient and competitive products [1, 2].

As the complexity and the competitiveness requirements of the products (in this case, mechanical and mechatronic systems) increase, the design and development times must be reduced, conditions in which the development and testing of physical prototypes become major impediments. Thus, it is necessary to implement design techniques based on modeling, simulation, and optimization in virtual environment, which can ensure a higher performance and quality of the products using only a fraction of the time and cost required in traditional approaches. Virtual prototyping is a computer-aided engineering-based discipline that entails modeling products, simulating and optimizing their behavior under real-world operating conditions. Through the use of various types of software solutions for evaluating the form, functionality, and durability of the products in an integrated approach, complex digital prototypes can be created and then used in virtual experiments (lab and field tests) in a similar way to the real cases [3–5].

In this context, the chapter proposes to present the integrated concept of modeling, simulation, and optimization of the behavior of mechanical and mechatronic systems through the use of a virtual prototyping software platform. The platform integrates specific software solutions for evaluating the form, assembly, functionality, and durability of mechanical and mechatronic systems. The components of the virtual prototyping platform are depicted by mentioning their particular role, as well as the mode in which they are integrated within the platform and communicate (data transfer) with each other. Then, the virtual prototyping stages are discussed starting from a flowchart reflecting the mode in which the data are transferred from one stage to another, to obtain a valid and optimal virtual prototype, these being the two attributes that the virtual prototype must have in order to be a truly useful/viable one. Finally, a case study is developed by considering a complex product, namely, a suspension system for motor vehicles, which is approached in mechatronic concept, by integrating the two main subsystems (the mechanical and actuating and control devices) at the virtual prototype level.

2. The software platform for virtual prototyping

In the general case, the virtual prototyping platform of the mechanical systems integrates three basic software solutions (**Figure 1a**): computer-aided design, multibody systems (MBS), and finite element analysis (FEA). In addition, in the case of mechatronic systems (mechanical systems with controlled actuation), the virtual prototyping platform integrates a design for control (DFC) software solution (**Figure 1b**), in the concurrent engineering concept (for the purpose of co-simulation).

Firstly, with the help of CAD software, the geometrical (solid) 3D model of the mechanical system is developed, with the purpose to determine the mass and inertia properties (moments and products of inertia) of the bodies (rigid parts). The 3D model is then transferred to the MBS software, which is intended to analyze and optimize the behavior of the mechanical system (in terms of kinematics, statics, and dynamics, by case). The data transfer from CAD to MBS is performed using

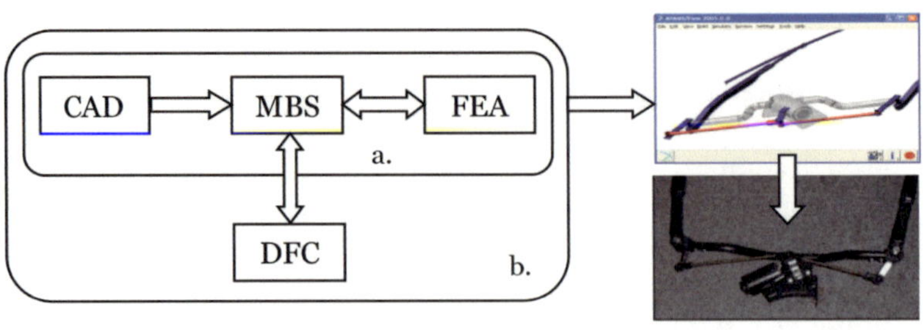

Figure 1.
The virtual prototyping software platform for mechanical (a) and mechatronic (b) systems.

standard geometry file formats, such as STEP, IGES, Parasolid, stereolithography, and others. From this point of view, there are no rules, but only certain recommendations of the software producers regarding the file format. For example, the recommended geometry transfer formats from the main CAD software to the MBS environment Automatic Dynamic Analysis of Mechanical Systems (ADAMS) of MSC Software (which is a global leader in virtual prototyping software and services) are presented in **Table 1**. With such file formats, the import into ADAMS is done through the general ADAMS/Exchange transfer interface. At the same time, specialized modules (interfaces) for geometry transfer were developed, which perform a customized transfer between the CAD and MBS ADAMS environments, as it is also presented in **Table 1** [6].

Initial Graphics Exchange Specification (IGES) format represents the first standard of interchangeability, being designed in American Standard Code for Information Interchange (ASCII) code. IGES reduces the CAD model to a list of entities, each entity being associated with a number. Drawing Exchange Format (DXF) is also based on graphical entities, for each data type, which is ASCII encoded, being allocated a line. Standard for the Exchange of Product Model Data (STEP) format describes the data at the product level and not the entity, through a specialized language (Express) that establishes the correspondence between the STEP file and the CAD model. Stereolithography (STL) format is a neutral format based on stereolithography, being used mainly in rapid prototyping devices (laser printing). Parasolid format is a geometric modeling kernel that allows transferring the entire 3D solid model through a single file, while in the case of the other formats, the transfer is done part by part (one file for each part).

The mass and inertia properties of the bodies are automatically calculated by the MBS software (ADAMS, in this case), depending on the 3D solid model imported from CAD and the associated material (defined by the well-known characteristics/properties: Young's modulus, Poisson's ratio, and density). Most MBS software solutions (including ADAMS) have their own solid modeling library, the modeling principles being the same as in CAD (elementary solids, composite solids using Boolean operations (union, extraction, and intersection), solids obtained by extrusion and, respectively, rotating surfaces), but for bodies with more complex geometry, the use of specialized CAD environment is required.

CAD software	File formats	Transfer interfaces
Unigraphics (UG)	Parasolid STL	UG/Mechanism
CATIA	STL STEP IGES	CAT/ADAMS
Pro/ENGINEER (currently Creo Elements/Pro)	STL IGES	MECHANISM/Pro
SOLIDWORKS	Parasolid STL IGES	Dynamic Designer
I-DEAS	STL IGES	Mechanism Design Mechanism Simulation
Mechanical Desktop	IGS STL DXF	Dynamic Designer

Table 1.
The file formats and transfer interfaces from CAD to ADAMS.

Based on the results of the dynamic analysis performed in the MBS environment, the system loads by forces and torques are determined, representing input data for the analysis with finite elements within the FEA software. Subsequently, the deformability state of the components is returned in the MBS software, thus making possible the dynamic analysis of the mechanical system with deformable (flexible) parts, which is more realistic (closer to reality) than the analysis with rigid parts [7, 8]. Through the analysis of compliant models, the stress and vibration states can be determined with the purpose to evaluate the functional and durability performances of the mechanical system. The data transfer from ADAMS to the main FEA environments (such as ANSYS, ABAQUS, or NASTRAN/PATRAN) is done through FEA Loads type format, from the general ADAMS/Exchange transfer interface. The FEA to MBS connection is usually made through the modal neutral file (MNF) format. In ADAMS, the data import from FEA is managed by the ADAMS/Flex interface. It should also be mentioned that ADAMS software package integrates a specialized module, called ADAMS/AutoFlex, which can be used for the conversion of rigid bodies into deformable equivalents, but with certain limitations in the case of more complex geometry bodies, for which it is still necessary to use specialized FEA software.

Finally, regarding the communication between ADAMS and the DFC software environments, in the case of mechatronic systems, the ADAMS package integrates the plug-in ADAMS/Controls through which the data transfer with one of the following DFC software is carried out: MATLAB/Simulink, $MATRIX_X$, and EASY5. Basically, ADAMS/Controls manages the input and output plants of the controlled process (as mentioned above, the outputs from the MBS are inputs into DFC and vice versa respectively), allowing to perform the co-simulation (in-parallel running/processing) of the two main subsystems of a mechatronic system, namely, the mechanical device and the actuating and control device. The information related to the input and output plants are saved in a specific file having the extension .m (for MATLAB) or .inf (for EASY5 and MATRIXX). At the same time, a command file (.cmd) and a data file (.adm) are generated, which are used during the co-simulation process. The files thus generated by ADAMS/Controls are then imported into the DFC application, where the ADAMS interface block is subsequently set and the control block model is designed. It should be mentioned that ADAMS provides some facilities for control system design, which are integrated into the Controls Toolkit module, but obviously not up to the level of complexity offered by dedicated DFC software.

By integrating the mechanical device and the control system at the virtual prototype level, the two models/subsystems are simultaneously tested and verified, thus simplifying the experimental testing process and eliminating (or at least minimizing) the risk that the control law is not accurately tracked (complied) by the mechanical device [9, 10]. Such a mechatronic concept approach is known as concurrent engineering. The simulation algorithm for mechatronic systems involves the following steps:

1. Within MBS software: modeling the mechanical device (including bodies, joints, actuating elements, other force generating elements), analyzing-simulating the MBS model, modeling the input (I) and output (O) plants in/from the MBS model, and exporting the MBS model for DFC

2. Within DFC software: importing the mechanical model, synthesizing the desired trajectories of the mechatronic system and modeling the input block diagram (reference signals synthesis), designing the control system block diagram, synthesizing the controller and the electrical interfacing circuits, and simulating the mechatronic system

The so described simulation process creates a closed loop, in which the controlled inputs of the control application affect the simulation in the MBS environment, while the outputs from the MBS simulation affect the level of the controlled signals in DFC.

3. The virtual prototyping process

A complete virtual prototyping process is defined by the following five stages (see also the workflow schematic representation in **Figure 2**): modeling, analysis, validation, refining, and optimization. During the modeling stage, the specific components of the mechanical or mechatronic system (such as bodies, connections between bodies, actuating elements, and other force generating elements) are created by using the software solutions shown in **Figure 1**. The output from modeling is the initial virtual model, which is then analyzed (simulated/tested) with the purpose to determine the behavior of the mechanical or mechatronic system, in terms of movement (linear or angular positions, velocities, and accelerations, by case) and reaction force states. The results obtained through the simulation in virtual environment (which are the analysis outputs) are then compared with the corresponding experimental results obtained by physical prototyping, in order to validate the virtual model. It should be mentioned that the physical prototyping is not a stage in itself of virtual prototyping, but a supporting process for this. By the comparative analysis of the virtual and experimental results, one of the following two cases can be reached: valid virtual model (when the virtual results fit with the experimental ones) and invalid virtual model (when the results obtained through the simulation in virtual environment do not match the experimental ones). In the first case, the last step of the virtual prototyping process will be the optimization, which aims to determine the optimal design of mechanical or mechatronic system (in terms of functionality, efficiency-energetic, or economic, by case). On the other hand, if the validation output is expressed by an invalid virtual model, the refining stage must be accomplished with the purpose to improve the fidelity of the virtual model by reference to the physical one. The refined virtual model is then analyzed (by simulation in virtual environment), followed by a new validation. In this way,

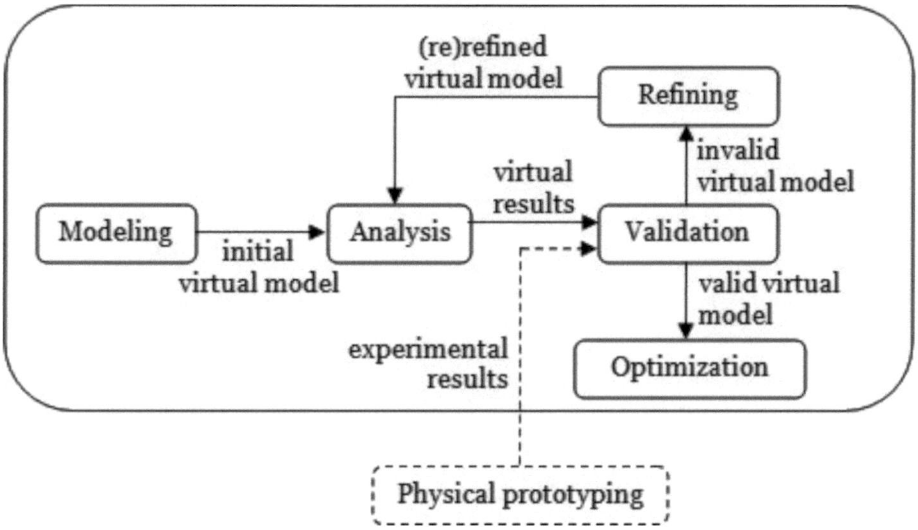

Figure 2.
The virtual prototyping workflow.

an iterative process (refining—analysis—validation) is carried out until a valid virtual prototype is obtained, which will be then the subject for optimization.

The basic principle for a successful virtual prototyping process can be formulated as follows: as complex as necessary and as simple as possible. This is in compliance with Einstein's statement: "A scientific theory should be as simple as possible, but no simpler." The idea is to manipulate the simplifying assumptions in a way that reduces the complexity of the virtual model (in order to make the real-time simulation), but without affecting/altering the precision of the results. In other words, a useful virtual prototype should be a trade-off between simplicity and realism. In the following, the implementation of this basic principle regarding the modeling and refining will be discussed for the basic components of a mechanical or mechatronic system, namely, bodies, connections between bodies, and actuating elements. For each of them, the real modeling case and the specific simplifying assumptions (hypotheses) are presented in **Table 2**.

In the real case, all the bodies are flexible (deformable), more or less, depending on the state of loading to which they are subjected, having constant mass (in most cases) and variable inertia properties (by changing the geometric shape). The simplifying assumptions for the modeling of the bodies are obtained from the real case by successively neglecting certain properties, as follows:

1. Rigid bodies: the shape of the bodies does not change during the analysis, so their inertial properties become constant.

2. Point masses: the shape is neglected by considering that the whole body mass is concentrated in a point (the center of mass), and in this way the inertia properties are not taken into account.

3. Bodies without mass (massless bodies): both the mass and the inertia properties are neglected as a consequence of the fact that the bodies are modeled by 2D elements/objects (such as lines, polylines, plane curves).

4. Composed restrictions: this is a special modeling case in which certain bodies are modeled as constraints between other bodies, such as constant distance or area constraints.

Components	Real case	Simplifying assumptions
Bodies	• Deformable (flexible) bodies	• Rigid bodies • Point masses • Bodies without mass (massless bodies) • Composed restrictions
Connections between bodies	• Deformable (flexible) contacts	• Rigid contacts • Joints • Constraint equations
Actuating elements	• Motors/actuators • Human operators • External factors	• Motor forces or torques • Motion restrictions

Table 2.
The modeling of the basic components.

The modeling of the bodies by composed restrictions is not possible for all the bodies in a mechanical or mechatronic system, but only in the following cases:

- The body is a mobile one, and not the fixed part of the system (which must remain the reference part to which the global reference frame is attached).

- The body is not input (by which movement is introduced into the system) or output part (from which the movement is collected), and this is because the movement can be introduced/collected only through/from bodies.

- No external forces or torques are applied to the body, or no force generating elements are connected to the body, and this is because the forces can only act on bodies.

A more detailed discussion on the modeling of the bodies as composed restrictions can be performed in correlation with the MBS models of the four-bar mechanism is schematically represented in **Figure 3**. So, the model shown in **Figure 3a** is a general one, with four bodies (three mobile parts, 1, 2, and 3, and the fixed part/ground, 0). Then, in **Figure 3b** and **c**, there models shown with three bodies (two mobile parts, 1 and 2 or 3, and the fixed part, 0) and one composed restriction (constant distance between the corresponding ends of the rod 2 and ground, respectively, of the crank 1 and rocker 3). Finally, in **Figure 3d**, the model is shown with a minimum number of bodies (the rod 2 and the fixed part) and two composed restrictions between the two bodies. The model with a minimum number of bodies is valid one only if the mobile body is at the same time input and output of the system, and this is possible because the body has three movements (two translations along the two axes of the representation plane and one rotation around the axis normal to this plane). The four MBS models shown in **Figure 3** are defined by the following numbers of generalized coordinates (movement parameters—6 per each mobile body): 18 (a), 12, (b and c), and 6 (d). Therefore, the model with a minimum number of bodies is the most convenient from the point of view of the complexity, which depends on the number of equations for determining the behavior of the system.

The connections between bodies (excepting the previously discussed composed restrictions, which are not connections with a physical equivalent) are nothing else than contacts between the geometric forms/shapes of the bodies. These contacts can be classified in two groups, with representative examples in **Figure 4**, as follows:

- Stationary (permanent) contacts (**Figure 4a**), where the connection between bodies is kept as in the initial state during the entire analysis range (throughout the system operation)

- Nonstationary contacts (**Figure 4b**), where the connection between bodies changes during the analysis, either the contact is lost or it occurs or the type of contact changes (e.g., from surface contact to linear or point contact)

Given that the real modeling case of the bodies is flexible (deformable) bodies, the real modeling case for the connections between bodies will be contacts between flexible bodies (or briefly, flexible contacts). Then, the first simplifying assumption for the modeling of the bodies (rigid bodies) is automatically transferred to the modeling of the connections, resulting rigid contacts (contacts between rigid

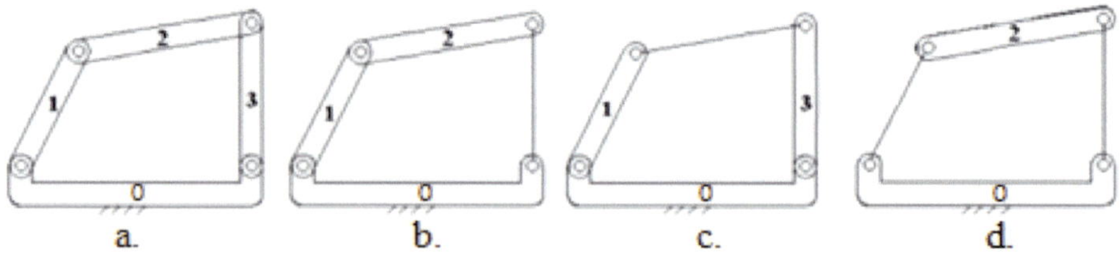

Figure 3.
MBS models for the four-bar mechanism: (a) 4-body model, (b, c) 3-body models and (d) 2-body model.

Figure 4.
Types of contacts (connections) between bodies: (a) stationary and (b) nonstationary.

bodies). Both in the case of flexible contacts and for rigid contacts, the connections do not restrict movements, but they introduce reaction forces and torques. The other simplifying assumptions for the modeling of the connections are the ones that restrict movements, namely, joints and constraint equations, which can be used only for the stationary connections (such as that shown in **Figure 4a**, where the contact between the two bodies of the hinge can be modeled as a revolute joint). It should be mentioned that the joint is a symbolical representation (like a modeling shortcut) in the software of the constraint equations, which can be classified in two categories: constraint equations generated by the software (through user-modeled joints), and constraint equations created by the user.

The actuating elements of the mechanical/mechatronic systems are usually found in the following categories (**Figure 5**): (a) rotary or linear motors/actuators, (b) external factors (such as the wind action for a wind turbine or the road irregularities for a vehicle suspension system), and (c) human operators. Whatever the case, the actuating elements generate mechanical power, which is defined by two components: force and movement. In these terms, the actuating element can be modeled by one of the two mentioned components: motor force/torque and motion restriction. The latter, by which the movement of the actuated (input) bodies is controlled, can be applied at the position, velocity or acceleration (linear or angular, by case) level, usually as in time variation laws. On the other hand, the motion restrictions can be applied in joints (thus controlling the relative motion between the adjacent bodies, as in the case of the jack mechanism shown in **Figure 5c**, where the relative motion between the crank and the fixed support can be controlled in the revolute joint between the two) or in points (thus controlling the spatial or planar positions of certain points of interest on the body, for example the point located in the end-effector extremity of the industrial robot shown in **Figure 5a**).

Figure 5.
Types of actuating elements of the mechanical/mechatronic systems: (a) motors, (b) external factor and (c) human operator.

4. The analysis flowchart

As it results from the ones presented in the second section of the paper, the central component of the virtual prototyping platform is the MBS software solution, which is the integrative solution used to simulate and optimize the behavior of the mechanical and mechatronic systems [11–15]. The analysis flowchart in MBS environment is schematically represented in **Figure 6**. The types of analysis can be performed separately or coupled in a certain sequence depending on the degree of freedom (DOF) of the mechanism, which expresses the number of uncontrolled (independent) movements, which take place under forces action. The basic components of a mechanical/mechatronic system can be structured in the following way: components that bring movement → mobile bodies; components that eliminate motion → connections between bodies (when they are modeled by joints or constraint equations); components that control motion → actuating elements (when they are modeled by motion restrictions). Thus, the number of degrees of freedom is given by the following equation (Gruebler's count) [16]:

$$DOF = 6n - \Sigma(r + r_m) \tag{1}$$

where n is the number of mobile bodies, Σr is the sum of geometric restrictions (joints and/or composed restrictions), and Σr_m is the sum of motion restrictions.

To better understand the above, **Figure 7** shows three modeling cases for an open-loop system formed by two bodies (the mobile part and the ground) connected by a revolute joint, with the following particularities: (a) there is no actuating element; (b) the actuating element is modeled by a motion restriction, controlling in this way the angular position of the rotating body/crank; and (c) the actuating element is modeled by a motor torque applied on the rotating body. For the three cases, the following numbers of degrees of freedom are corresponding: (a) DOF = 6–5 = 1; (b) DOF = 6 – (5 + 1) = 0; (c) DOF = 6 – 5 = 1. Therefore, the second model (b) has no independent motion, the angular positions of the crank being controlled (imposed) by the motion restriction regardless of the mass and the inertial properties of the body. In the first (a) and the third (c) case, respectively, the model has one uncontrolled motion (the rotation of the crank), which is influenced by the action of the forces (mass and inertia forces in the both cases, and in addition the motor torque in the third case). Thus, the motion restrictions remove degrees of freedom, by controlling the motion, while the motor forces/torques do not remove degrees of freedom.

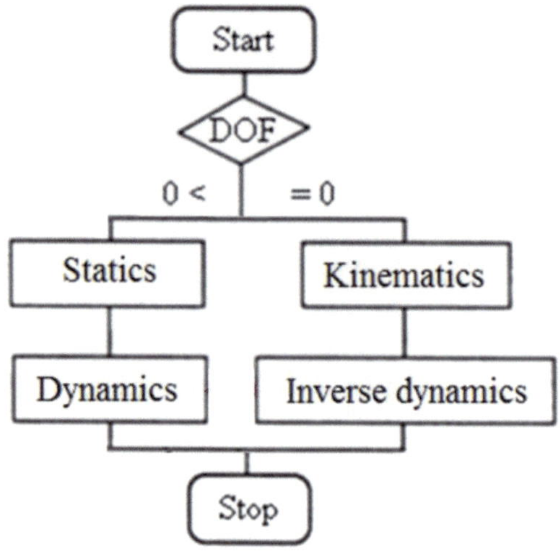

Figure 6.
Analysis flowchart of the mechanical/mechatronic systems.

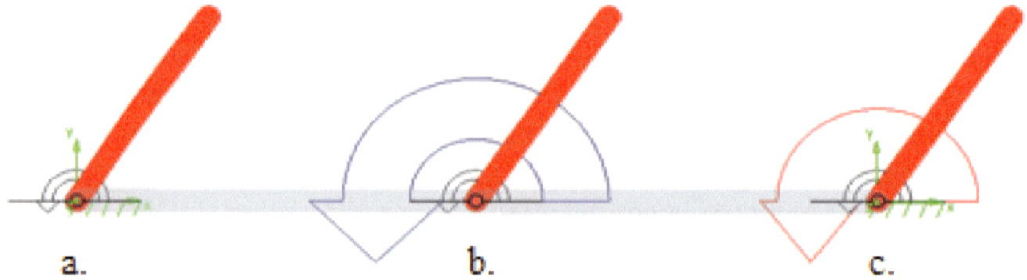

Figure 7.
(b) Controlled vs. (a, c) uncontrolled movement.

The four types of analysis shown in **Figure 6** are defined by the following:

- Dynamics: inputs, the assembled configuration (bodies and connections), and the loads through forces and/or torques (all of them) outputs, the time histories of motion and reaction states

- Kinematics: inputs, the assembled configuration and motion restrictions (no forces/torques), and outputs, the time histories of motion

- Statics: inputs, the assembled configuration and the loads through forces and/or torques (excepting the forces that depend on velocity and acceleration, such as damping and inertia forces), and output, the equilibrium configuration

- Inverse dynamics: inputs, the same as in dynamics, but with the actuating elements as in kinematics, and outputs, the motor forces/torques

Considering the particularities of the simplifying assumptions for the modeling or refining of the basic components of the mechanical/mechatronic systems (as presented in the 3rd section of the paper) and those of the types of analysis mentioned above, **Table 3** shows the correlations between the simplifying assumptions and the analyses, which can be interpreted as validity fields for hypotheses (i.e., analyzss where the use of hypotheses does not generate errors).

Components	Simplifying assumption	Analysis
Bodies	Rigid bodies	Dynamics
		Inverse dynamics
	Point masses	Statics
	Massless bodies	Kinematics
	Composed restrictions	
Connections between bodies	Rigid contacts	Dynamics
		Inverse dynamics
	Joints/constraint equations	Kinematics
		Statics
		Dynamics
		Inverse dynamics
Actuating elements	Motion restrictions	Kinematics
		Inverse dynamics
	Motor forces/torques	Dynamics
		Statics

Table 3.
The validity fields of the simplifying assumptions.

The analysis methodology of the mechanical/mechatronic systems by using MBS software environment (ADAMS in this case) involves three stages: pre-processing (system modeling), processing (model running), and post-processing (processing results). The pre-processing stage involves indicating the input data, as follows: specifying information regarding the calculations to be performed, such as the type of analysis to be carried out, the units of measurement, the type of coordinate system (e.g., Cartesian), the gravitational acceleration vector, and the analysis time interval and modeling the components of the mechanical/mechatronic system (bodies, connections between bodies, actuating elements, and other force generating elements, such as springs or dampers, by case). The processing stage is performed automatically by the program, and consists from generating and solving the algebraic and differential equations that mathematically describe the system. The post-processing stage consists of processing the analysis results, by drawing variation diagrams/charts, generating tables with numerical values, and creating graphical animations, all of which providing an overview of how the mechanical/mechatronic system behaves.

5. Case study

Based on the above, a case study corresponding to a high complexity system, namely, a suspension system for motor vehicles, is presented below. The virtual model contains the front and rear wheel suspension subsystems, as well as the actuating subsystem. The prototype is used for simulating the passing over bumps dynamic regime under laboratory conditions, through the use of a virtual testing bench (**Figure 8**). The approach is a mechatronic one, in the sense that the actuating subsystem, containing four linear actuators that sustain/support and move the wheels, is controlled in such a way as to ensure the desired running path profile by the vertical displacements that apply to the wheels. The virtual model of the mechanical device (including the two suspension subsystems, the car body, and the testing bench) was developed by using the MBS software environment ADAMS. The 3D solid model was developed with the help of the CAD software environment CATIA, the transfer to ADAMS being performed as described in the second section of the chapter.

The front wheels suspension subsystem (**Figure 9**) contains two independent Short-Long Arm (SLA) mechanisms, also called double-wishbone. The lower (long) and upper (short) arms of the mechanism are double-hinged to the car body by using bushings (compliant joints), while the other ends (outward) of the arms are connected to the wheel carriers by spherical joints. The same type of joints was used for the connections of the steering rods to the adjacent parts (wheel carriers and car body). The rear wheels suspension subsystem (**Figure 10**) ensures the guiding of the whole axle in the relative movement to the car body by a so-called 4S

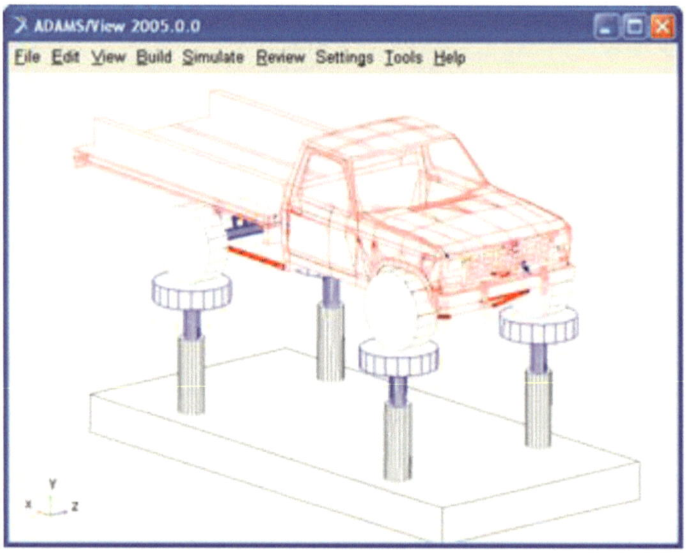

Figure 8.
The MBS virtual model of the vehicle.

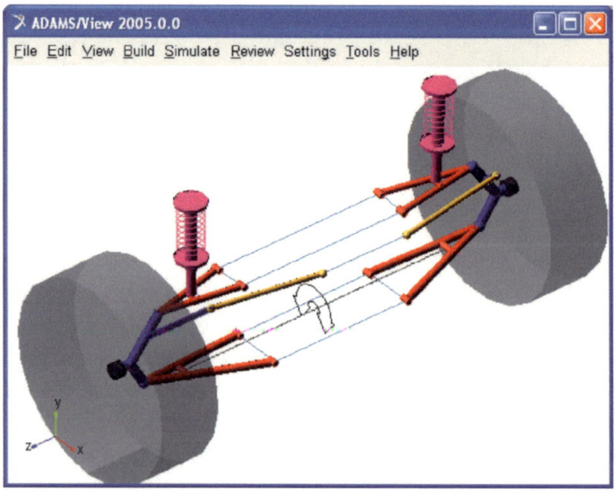

Figure 9.
The front wheels suspension subsystem.

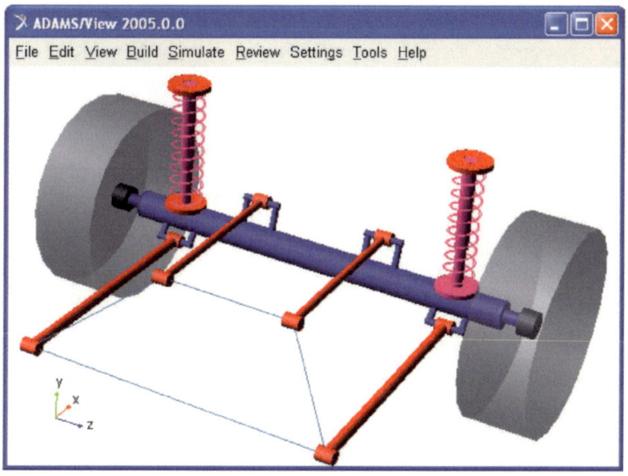

Figure 10.
The rear wheels (axle) suspension subsystem.

suspension mechanism, with four longitudinal arms that are connected to the adjacent bodies by bushings. In the case of the front suspension, the spring and damper groups are mounted between upper arms and car body, while for the rear suspension, these elastic and damping elements are arranged between axle and car body. For the both suspension systems, the bumpers limiting the run (extension, and respectively compression), which are nonstationary elastic elements, are disposed inside the dampers, thus limiting the relative displacement between the two parts of the damper (cylinder and piston).

The connections between the wheels and the upper platters of the actuators were modeled by contact forces between the corresponding geometries, which allow considering the stiffness and damping properties of the tires, as well as the friction between the bodies. As mentioned, the vertical displacement of the actuator plates is controlled so as to simulate the passing of the vehicle over various types of obstacles (road irregularities), the movement being transmitted to the wheels and then, through the suspension mechanisms, to the car body. The control system for the actuating elements (**Figure 11**) was designed with the help of the DFC software solution engineering analysis systems (EASY5). In this model, the MBS mechanical device is referred by the ADAMS Mechanism block. It should be mentioned that ADAMS and EASY5 software solutions are produced—marketed by MSC Software Corp., so the compatibility between them (in terms of facilities for data transfer for the purpose of co-simulation) is very good.

The modeling of the actuation and control system was carried out in mechatronic concept, by integrating the mechanical model (**Figures 8–10**) and the control system model (**Figure 11**) at the virtual prototype level. Thus, the two models (MBS and DFC) are being tested—verified simultaneously, minimizing the risk that the control law not to be followed by the mechanical model. In this case, the mechanical and control models are connected and communicate one with other through the use of ADAMS/Controls. The communication scheme between the MBS model and the control system is shown in **Figure 12**. The inputs into the mechanical model (outputs from the control system) are the motor forces developed by the four linear actuators, while the outputs from ADAMS (inputs into EASY5) are the vertical positions of the wheel actuator plates.

The input and output plants were defined by using a set of ADAMS state variables. The input state variables (the motor forces) are defined in ADAMS by null values, going to receive their values from the control application. The input variable

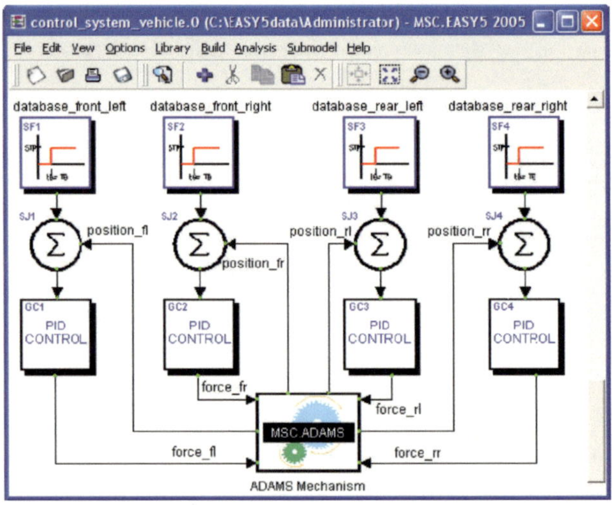

Figure 11.
The DFC model of the control system.

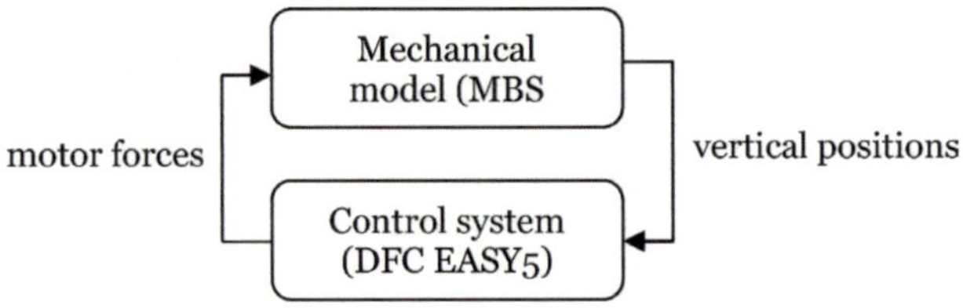

Figure 12.
The input and output plants.

is called by using the predefined function VARVAL (variable), which returns the value of the variable. For the output state variables, the time functions return the linear displacements along the vertical axis (Y). The output variables are modeled by using the predefined function DY (To Marker, From Marker), where the markers represent coordinate systems belonging to the adjacent parts (actuator cylinder and piston respectively), placed in the translational joint between the two parts of the actuator. The input variables are reported by plant input, PIN1–4, and the output variables by plant output, POU1–4. Information related to the input and output plants are saved in a specific file for EASY5 (*.inf); at the same time, a command file (*.cmd) and a data file (*.adm) are generated for the subsequent co-simulation. In the ADAMS Mechanism interface block, the execution mode is then defined; in this case co-simulation, specifying also the interval with which ADAMS/Controls, writes the results to files and adapts the animation and the communication range between ADAMS and EASY5 [17].

As mentioned, the vertical positions of the wheel actuator plates are imposed to simulate the passing of the wheels over bumps (road irregularities). For the study presented in this work, it was considered the road profile shown in **Figure 13**, which includes four speed bumps with a height of 20 mm. The delays between the excitations of the wheels (front-rear and left-right, respectively) correspond to a vehicle speed of 20 km/h (the vehicle has the wheelbase, i.e., the distance between the front and rear axles, of 2.4 m). In the DFC (control system) model shown in **Figure 11**, the imposed movement laws were defined by using the step function generator blocks SF1–SF4. This type of input data block is defined by the time at initiation of step input (To_SF) and the step input value (STP_SF). The step is triggered when time equals T0_SF and steps to a value of STP_SF, in accordance with the following conditional function:

$$\text{IF(TIME < T0\_SF) THEN S\_Out\_SF = 0 ELSE S\_Out\_SF = STP\_SF,} \tag{2}$$

where S_Out_SF is the step signal output, which is compared (by using a summing junction block) with the current position provided by the MBS model.

For each of the four linear actuators, PID controller was used as a control element. This controller corrects the difference between the imposed (desired) and current (measured) values of a specific parameter by computing and applying a compensatory measure that adapts the system properly [18, 19]. The optimal design of the controller can be achieved both by methods specific to the control theory (e.g., root locus, frequency methods), as well as by optimal parametric design techniques [20–22]. For this work, the optimization was performed by using the scripting capabilities integrated in EASY5 Matrix Algebra Tool (MAT), the control system model being managed as an EMX function. The optimization procedure is similar with that presented in [23]. The optimization goal is to minimize the difference between the imposed and current values of the actuator plate vertical position, while the design variables are the proportional (P), integral, (I) and derivative (D) factors of the controller.

In the conditions specified above, the time history variations of the vertical positions of the front and rear actuator plates are shown in **Figure 14**. The mechanical powers developed by the four linear actuators for generating the predefined

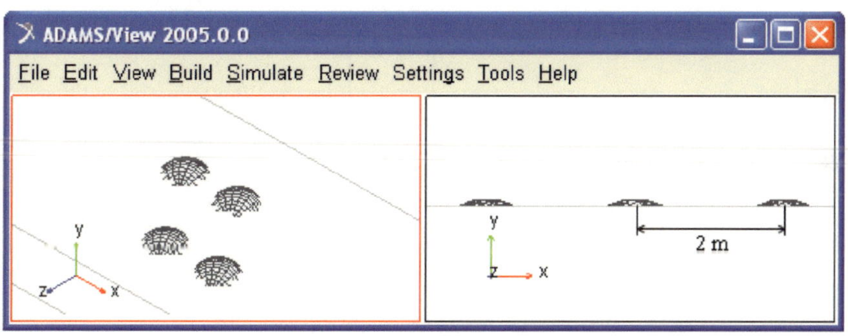

Figure 13.
The road profile simulated by the virtual test bench.

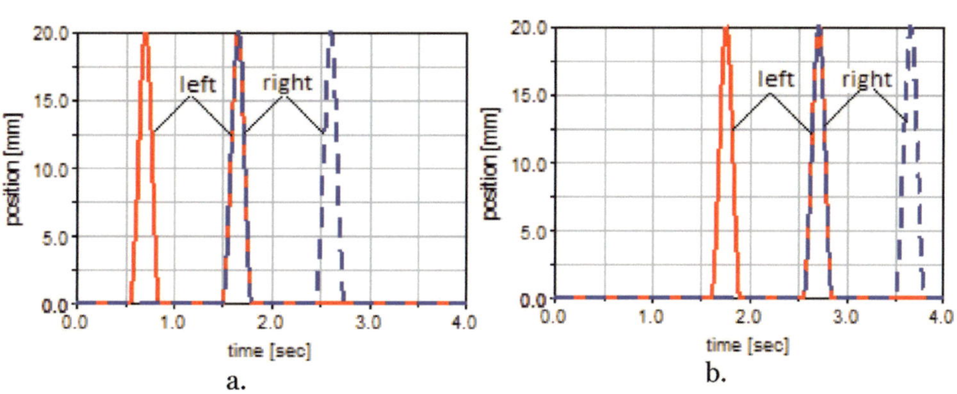

Figure 14.
The vertical displacements of the front (a) and rear (b) actuator plates.

movement laws, which were determined by multiplying the motor forces by the linear velocities of the actuator plates, are the ones shown in **Figure 15**. Some results that describe the dynamic behavior of the vehicle are presented in

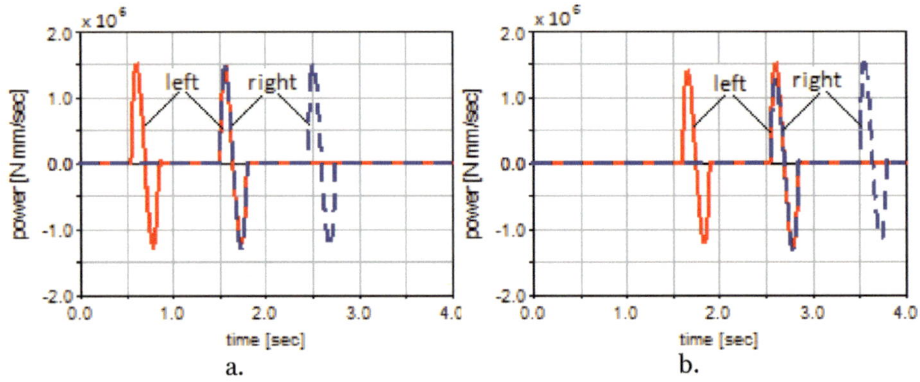

Figure 15.
The power developed by the front (a) and rear (b) linear actuators.

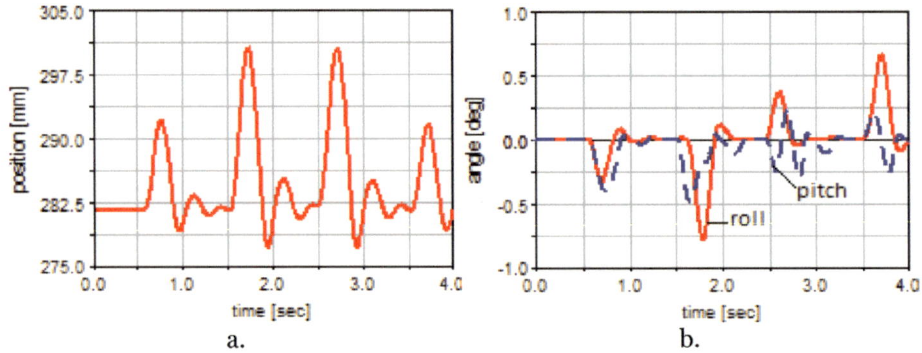

Figure 16.
The main linear (a) and angular (b) oscillations of the car body.

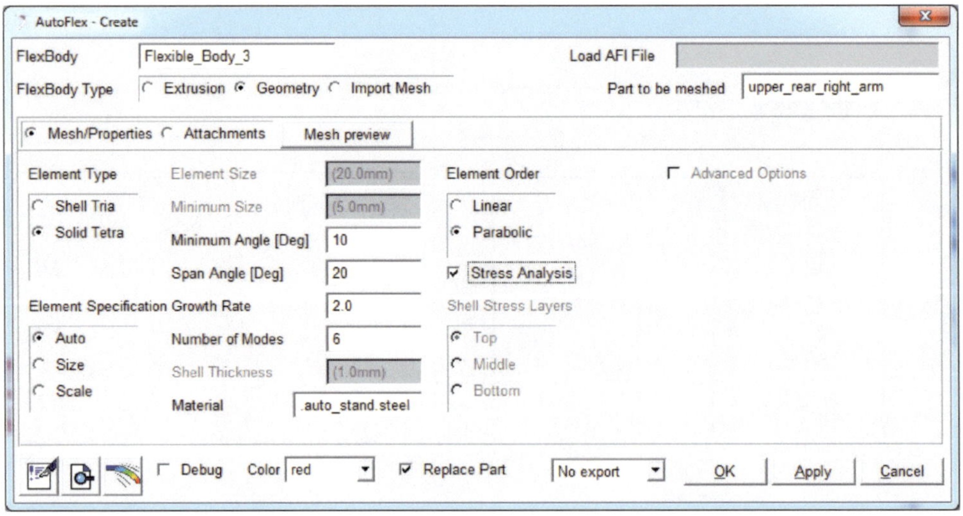

Figure 17.
Example of conversion from solid body to flexible body.

Figure 16, namely, the vertical displacement of the car body (a), which is measured in its center of mass and the roll and pitch oscillations/angles (b).

Further, the guiding arms of the front and rear suspension mechanisms, which were initially modeled as rigid bodies, were discretized into finite elements, for studying their deformability and stress state. The conversion from solid to flexible was achieved by using ADAMS/AutoFlex. For example, **Figure 17** shows the conversion window for the rear upper right arm, while in **Figure 18** the finite element model of this body, along with its first three vibration modes. It should be mentioned that ADAMS/AutoFlex allows viewing 18 vibration modes per flexible body, each mode of vibration being characterized by a modal frequency and a mode shape [24]. A dynamic simulation graphical frame for the compliant model (with flexible bodies), focused on the rear axle guiding mechanism, is shown in **Figure 19**, revealing the stress state of the guiding arms.

Many other results can be extracted from the simulations in virtual environment, for all the objects—components of the virtual model (e.g., bodies, connections between bodies, actuating elements, elastic and damping elements) and for any type of parameter (e.g., motion, force, energy), including results that cannot be measured experimentally for various reasons (such as lack of adequate sensors, hard to reach areas, high temperatures in the measuring area, and others). At the same time, by studying the influence of the various parameters that define the model (such as the global coordinates of the joint locations or the elastic and damping coefficients of the spring and damper assemblies) on the vehicle behavior, its kinematic and dynamic optimization can be simplified, by selecting the parameters which significantly influences the comfort, stability, or maneuverability of the vehicle.

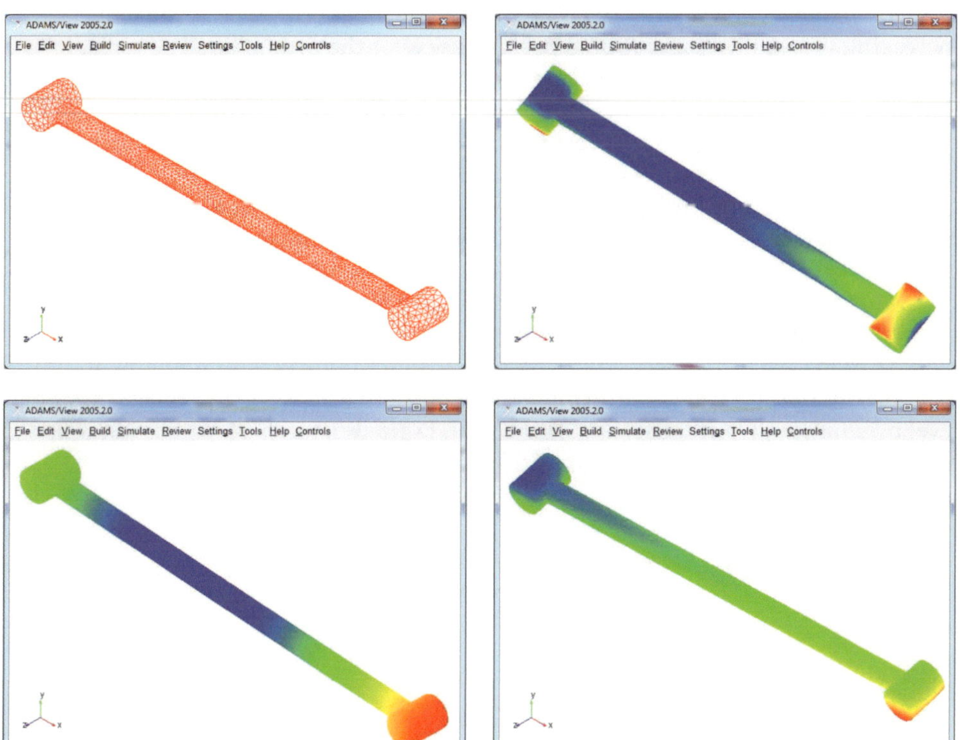

Figure 18.
The FEA model of the rear upper right arm.

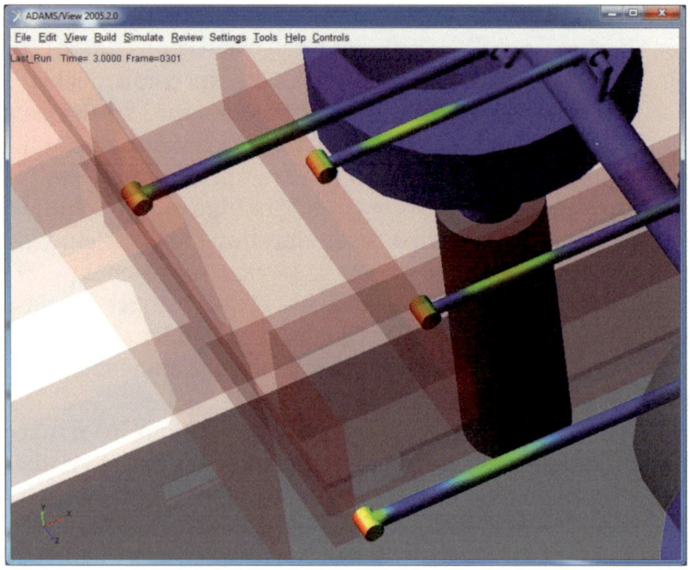

Figure 19.
Dynamic simulation graphical frame.

6. Conclusions

The use of virtual prototyping software platforms in the analysis and optimization of the mechanical and mechatronic systems offers important benefits, which focus on reducing the costs, as well as the design and development time while increasing the quality (operational performances of the products). The virtual prototypes are not made from real materials (such as steel, aluminum or wood) that are generally expensive, but from bits, with which any type of material can be simulated. Other significant cost reductions result from the fact that virtual prototyping does not involve destroying prototypes during testing (e.g., in the real car crash tests), the virtual prototype being restored to its original state by a simple mouse click. Multiple design variations (in various parameter combinations) can be explored early, without going through expensive (and often superficial) physical prototyping cycles. The virtual prototyping technique allows the replication on computer of both the product itself and the specific operating (working) environment. Among the critical success factors regarding the successful implementation of the virtual prototyping platforms, we can point to well-defined process, system-level orientation, efficient setting of the goal, rapid dynamics of the simulation, and high-quality infrastructure (hardware and software).

Author details

Cătălin Alexandru
Transilvania University of Brașov, Romania

*Address all correspondence to: calex@unitbv.ro

References

[1] Alexandru C, editor. Modeling and Simulation in Engineering. Rijeka: InTech Open; 2012. p. 298. DOI: 10.5772/1415

[2] Ryan R. Functional Virtual Prototyping - Realization of the Digital Car. Ann Arbor: Mechanical Dynamics Inc.; 2001

[3] Bernard A. Virtual engineering: Methods and tools. Journal of Engineering Manufacture. 2005;**219**(5): 413-421

[4] Haug EJ, Choi KK, Kuhl JG, Vargo JD. Virtual prototyping simulation for design of mechanical systems. Journal of Mechanical Design. 1995;**117**(63):63-70

[5] Mejia-Gutierrez R, Carvajal-Arango R. Design verification through virtual prototyping techniques based on systems engineering. Research in Engineering Design. 2017;**28**(4):477-494

[6] Getting Started Using ADAMS/ View. Newport Beach: MSC Software Corporation; 2005

[7] Ambrósio JA, Gonçalves JP. Complex flexible multi-body systems with application to vehicle dynamics. Multibody System Dynamics. 2001;**6**:163-182

[8] Da Silva MM, Costa NA. Handling analysis of a light commercial vehicle considering the frame flexibility. International Review of Mechanical Engineering (IREME). 2007;**1**(4):334-339

[9] Alexandru C, Pozna C. The virtual prototyping of the windshield wiper systems in mechatronic concept. In: Proceedings of the 11-th European Automotive Congress (EAEC '07). Budapest: Scientific Society of Mechanical Engineers; 2007. pp. 58-69

[10] Enescu M, Alexandru C. Virtual prototyping of a spraying robotic system. Environmental Engineering and Management Journal. 2011;**10**(8):1197-1205

[11] Fischer E. Standard multi-body system software in the vehicle development process. Journal of Multi-body Dynamics. 2007;**221**(1):13-20

[12] Garcia de Jalón J, Bayo E. Kinematic and Dynamic Simulation of Multi-Body Systems. 1st ed. New York: Springer - Verlag; 1994. p. 440

[13] Haug EJ. Computer Aided Kinematics and Dynamics of Mechanical Systems. 1st ed. Needham Heights, Massachusetts: Allyn & Bacon; 1989. p. 500

[14] Schiehlen WO. Multi-body systems dynamics: Roots & perspectives. Multibody System Dynamics. 1997;**1**(2):149-188

[15] Shabana A. Dynamics of Multi-Body Systems. 4th ed. New York: Cambridge University Press; 2014. p. 393. DOI: 10.1017/CBO9781107337213

[16] Orlandea N. Design of parallel kinematic systems using the planar enveloping algorithm and ADAMS program. Journal of Multi-body Dynamics. 2004;**218**(4):211-221

[17] EASY5 Guide Help. Newport Beach: MSC Software Corporation; 2005

[18] Dorf RC, Bishop RH. Modern Control Systems. Boston: Pearson; 2016. p. 1032

[19] Haque HH, Hassan HM, Hossain SM. Comparison of control system using PLC & PID. In: Proceedings of the ASEE 2014 Conference. Bridgeport: American

Society for Engineering Education; 2014. p. 99(1-6)

[20] Alexandru C. Optimal design of the controller for a photovoltaic tracking system using parametric techniques. Annals of the Oradea University: Fascicle Management and Technological Engineering. 2010;**IX**(1):2.1-2.9

[21] Alexandru C. Optimal design of the mechanical systems using parametric technique & MBS (multi-body systems) software. Advanced Materials Research. 2012;**463-464**:1129-1132

[22] Alexandru C. Optimal design of the dual-axis tracking system used for a PV string platform. Journal of Renewable and Sustainable Energy. 2019;**11**(4):043501(1-14)

[23] Alexandru C, Alexandru P. Control strategy for an active suspension system. World Academy of Science, Engineering and Technology. 2011;**79**:126-131

[24] Blevins RD. Formulas for Natural Frequency and Mode Shape. Riverside, Connetticut: Krieger Pub; 2001. p. 506

Chapter 5

Renovation and Reuse of Waste Electrical and Electronic Equipment in the Direction of Eco-Design

Panagiotis Sinioros, Abas Amir Haidari, Nikolaos Manousakis, Michael Lasithiotakis and Ourania Tzoraki

Abstract

Nowadays there is a higher need of strict and broader legislation in waste electrical and electronic equipment (WEEE) recycling industry to reduce environmental effects of WEEE. Environmental challenges include pollution, exhaustion of natural resources, waste management and reduction of landfills. High speed in technological development in many sectors puts many products in great challenge of obsoleting almost immediately after their purchase. In particular, this is the fate for electrical and electronic equipment (EEE). They are forever-improving and incorporate state of the art innovations. This provide many benefits; however, at the same time, its expansion results in rapidly growing waste stream of WEEE. WEEE contains a combination of all these situations, including for example, batteries, plastics of quality, precious metals and toxic soldering metals. The reuse and renovation of WEEE are therefore very critical because of its significant ecological environmental impacts. Sustainable development is not a static situation, but a state of dynamic balance between human and environmental system. The current chapter explores sustainability planning and strategies such as eco-design, and design for dismantling and recycling, and what they mean for electronic products. It examines the incentives, methods and tools for sustainable electronic product design, with particular emphasis on reuse, recycling, selection of sustainable materials and processes, and lack of resources.

Keywords: WEEE, eco-design, reuse, recycling, DFS, reconstruction, renovation, repair, life cycle

1. Sustainable construction, planning and development and reuse procedures: design for sustainability

WEEE production is growing every year [1]. However, the product life cycle analysis demonstrates that disposal phase contributes substantially on the environmental impacts of WEEE [2–4], especially in products containing toxic materials, rare or valuable materials, or materials with high energy content. The world's current experience of financial crisis and climate is a major crisis nowadays that seems to be linked. With implementation of improved regulatory and control mechanisms, avoiding future financial crisis is possible, while scientists predict no regulation will save the planet from devastating consequences. Climate change further than environmental issues has serious social implications, from displacing people from areas they lived for generation to rising food prices. In addition, it will create economic threats in many countries that are comparable in size to world wars [5].

There is a third serious crisis that we are already facing: the exhaustion of limited natural resources. Finishing mineral oils is well known, but other natural resources too, for example, rare earths and precious metals. Nowadays, the objectives of wars and conflicts are land, water, food and mineral resources, and it is a proof that production and consumption have reached saturation point that cannot be a model of growth for a planet that will reach 9 billion in 2050. Despite efforts, industrialized countries consume 70% of global resources and host only 20% of world's population. Three sectors of consumption are primarily responsible for this: food/agriculture, transport/tourism, housing/energy consumption in buildings. These sectors account for about 80% of environmental impact of EU countries [6].

Brundtland Commission formulated "Sustainable Development" model in 1987 [7] as a development that meets needs of present living generations, without compressing the ability of future generations to meet their needs. In 1992, more than 170 countries agreed to fight for sustainable development as set out in Agenda 21 [8], where specific work on production, consumption and policy is formulated, and possible meters are proposed. Despite the breadth and complexity of the issue, these six key principles describe how a viable community should interact with other communities and nature:

- Protection of the environment: Protection of resources and life support systems needed to maintain human well-being and life.

- Development: Improving "quality of life," whose economic growth is a part of it, not a single objective!

- Future: It takes into account the interests of future generations in relation to what we leave behind.

- Equality: Fair distribution of resources, between developed, developing and even least developing countries (as each has a role to play and a cost to pay) in transition to sustainability. In particular, the most vulnerable (least developed countries) should be top priorities.

- Diversity: Different environmental, social and economic systems are generally more powerful and less vulnerable to irreversible or catastrophic damage. It is also a choice to more sustainable choices.

- Participation: Sustainability is not imposable; it requires support and involvement of all society's sectors and communities. This requires ensuring opportunities for participation in decision-making process.

1.1 Characteristics of sustainable development

- Conservation of resources

- Respect for the views of all interested parties

- Cooperation and partnership

- Follow-up to the precautionary principle
- Encourage subsidiarity: make decisions at the lowest achievable level
- Promoting personal freedom: satisfying needs without harming the environment or people
- Esthetic treatment: protect and create beauty spots and items

The term "sustainability" has been introduced by Victor Papanek in his publications [9, 10]. Already significant networks, including the environmentally friendly-designing international O_2 network (www. o2.org), established as early as 1988 [11] and, even more lately, the biggest networks and design establishments, have been effecting to introduce sustainability-design practices (DFS). Although there is still a lot of surface debate and some initiatives, especially from big companies, can be detected as green "laundering" (false or excessive green claims in advertising without real action), the current trend with DFS is constantly increasing.

With small efforts in research, it is apparently lagging behind in current development and market demands of DFS professions, and the few available programs for the growing number of young and enthusiastic students who they want to engage in DFS. Designers in broadest sense, including engineers and commercial creators, are still very often part of the current sovereign economic system, aiming at quantitative development as the sole objective of encouraging growing consumerism and wastage due to disposable products. These cause massive flows of resources from nature to waste disposal areas within a shockingly short period of time and the sale, through advertising and communication, of "goods" that you do not need but promote a modern useless lifestyle as the only desirable model of world prosperity.

Furthermore, engineers and scientists are not often strategic decision-makers and operate just at the bottom of the command structure. Sustainable development managers need to sit at the decision-making desks to push a real change. They need to be "loaded" with expertise of sustainability-based research and evidence, but also methods for evaluating and directing development and evidence-based planning decisions. They need to know about the history, problems and motivations of DFS theory and practice, and should adopt a more participatory practical design by first listening to stakeholders, understanding their problems and motivations, and then trying to develop more sustainable solutions [12].

2. Sustainability and eco-design of EEE with reuse of parts and materials

2.1 Reuse of parts and materials

Generic reuse strategies (include recycling, repair, rehabilitation and rebuilding) all are important strategies for sustainable production, because they help reduce landfill and the need for new material to be used in production. Rebuild, reconstruction and restoration (also referred to as recycling of products, commodity recuperation or resale market functions) are the various manufacturing techniques that use elements from used materials and are advantageous for recycling (recovery/reuse of content). Recycling defines the number of activities that gather, identify, process and use recycled materials in the manufacture of new products [13]. The advantages of restoration to recycling typically involve the following [14]:

- Reconstruction is a practice of "addition," whereas recycling is just a method of "lowering," so restoration of the material increases the value to the waste by converting it to functional condition. But at the other hand, recycling decreases the commodity to its raw resources.

- Reduction (which was) used in the manufacturing of the original material will be wasted after recycling it. The objective of this is to restore as complete as feasible the consumer goods, thus keeping energy and incoming resources of their first construction. However, recycling loses most of this energy and resources through reduction in raw materials. And this loss is even greater if energy used to extract raw materials and transport them is taken into account.

- Using the raw material recovery strategy, the waste of energy and assets to extract a marketable product from the waste product is greater. This may be because twice resources are expended on the processing of raw materials. Essentially, the product is "reduced" to resources (e.g., pouring smelting) but also, afterward, the resurrected materials are turned into goods.

- Engineers might be hesitant to using recycled materials since their performance could be uncertain [15]. Reconstruction is typically much more efficient in bulk material restoration than recycling, particularly for structural and electromechanical massive and complex materials.

The decision to implement product recovery should be carefully considered, as it may in some cases be counterproductive to sustainable development, for example by helping ineffective products to remain in circulation longer than is desirable. This is the case when a newer generation of products tends to be more environmentally friendly and efficient in operation; for example, new technology washing machines usually require less water, detergent and electricity. It is better that ideal product recovery is being used when it is both profitable and beneficial to the environment. Other issues to consider include creating new business models that include an effective reverse logistics system and ensuring sufficient quantities of used products (cores) to support product recovery processes. This is particularly important as consumers will only buy recovered products if they are considerably less expensive than other new products [16]. In the example of residential commodities, guaranteeing sufficient distribution of product is especially challenging, as it cannot be defined when these consumer goods will complete their lifespan. Optimally, commodity recuperation operations must be focused on eliminating substantial greenhouse emissions from the conveyance, as sections of the cycle actually occur at separate locations or, worse, when the waste material is shipped for processing and afterward brought back for purchase into the home country. There is a ranking predominantly centered on quality within a material recuperation method. Restoration is at the peak of this ranking since it's the only method of commodity recuperation that can return discarded goods in terms of quality, reliability and consistency to a standard competitive to that of new alternative commodities.

2.2 Sustainability and eco-design of EEE

Sustainability and eco-design of electrical and electronic equipment can be achieved by consumer awareness and the search for sustainable solutions, the requirement for legislation that ensures greater producer responsibility and green-sustainable programs, ensuring a competitive advantage for companies leading such efforts, the development of social responsibility and the need for business owners and consumers to invest their money and whatever that is reasonable and useful for people and the planet.

The EU's most important legislative instruments on the design of electrical and electronic equipment aim to save energy, manage the end of life and resource efficiency as key issues. The key legislative acts are Directive 2002/96/ EC on WEEE and Directive 2002/95/EC on the restriction of the use of certain hazardous substances in electrical and electronic equipment (RoHS). The WEEE and RoHS Directives are indicative of a policy strategy of boosting manufacturer accountability for recycling and management funding, reduce the environmental impact of electrical and electronic equipment and promote the recycling of valuable resources [17].

In conjunction with both the WEEE and RoHS Directives, Directive 2005/32/EC [18] laying down a guideline for establishing environmentally friendly design requirements for EUP is also significant because it spans the entire life cycle of energy-based products and the establishment of particular environmentally friendly design requirements for specific product groups through stakeholders involved in consultation processes. After end-of-life issues have been addressed with the WEEE and RoHS Directives, environmental significance of use phase due to energy consumption has also become evident in relation to increase in climate change threats. Thus, EUP Directive predicts energy efficiency of electrical and electronic products. However, additional design requirements for certain products (e.g., dishwashers and washing machines) that consume water and detergents during use phase are addressed in EUP reports and documents [19] and items with an impact on power consumption namely insulation, window frames, waterproof illumination, etc. The important regulatory platform is the current waste class division implemented in the European Waste Directive, which encourages waste reduction and preservation and whether products and resources are circulated. This would also influence design specifications, such as easy access to useful materials and competitive recycling capability, including recycling technologies. That package of all these four guidelines may be regarded as an attempt to achieve the European Integrated Policies Policy (IPP) goals.

Strong policy instruments to support sustainable planning activities also include the so-called green public procurement (GPP) or sustainable contracts. The public sector typically accounts for 10–20% of GDP, and annual investment from public procurement alone in the EU amounts to 72 trillion or 17% of GDP [20]. The "green procurement" has a growing importance in Europe. However, its pace of implementation varies considerably between various European countries. Current developments in EU policy and the EU 2020 strategy [21] further strengthen ambitions in the field of green public procurement and their implementation in practice. Most EU Member States have adopted national action plans for sustainable procurement, including objectives and implementing measures. An important factor in eco-design is ecological and communication labels, which remains a useful tool for rapidly exchanging information along the supply chain and between consumers. The most eco-labels for electronic products include the Nordic Swan, the Blue Angel (Blaue Engel) and the American Energy Star Label, the EU eco-label or the various energy efficiency labels defined in the EU Member States for specific product categories, such as large household appliances [22].

Simultaneously, policymakers, manufacturing firms, customer groups, journalists and also investors are increasingly seeking data from end-users, manufacturers and retailers regarding environmental consequences and sustainable development. Therefore, it is very significant to be able to track the products and their components to the original source throughout the entire supply chain. It is compulsory for those who import consumables into the EU market and their suppliers to provide data on adherence to EUP specifications. Across the USA, government requires corporations to use industry consultants to guarantee that their raw materials are not troublesome [23]. The problematic metals are widely used in electrical and electronic equipment (e.g., tantalum, gold, tungsten and derivatives). This requires the management of information along the supply chain, the standardization and digitization of relevant information, and respect for the confidentiality and competitive advantage of this information.

With regard to the standardization of EEE products, many national and international standards relate to sustainable design. International Organization for Standardization (ISO) published in 2002 a technical report ISO/TR 14062 [24]. This report refers to methods, tools and best practices for integrating environmental aspects into product design and development. It is currently being transported to ISO 14006 for the implementation of eco-design in Environmental Management Systems under the ISO 14000 series (ISO 14006 [25]). Although neither the technical report nor the new standard is intended to be used as standards for certification and recording purposes, certification bodies and companies have already used them for labeling activities, for example, environmental statements for cars, such as KIA [26] and Daimler [27]. As consumer demand for sustainable products is constantly increasing, new research bodies with ecological and social consciousness, for example, the Sinus Institute Heidelberg [28], describe consumer behavior.

3. Comparing reuse options; reconstruction, renovation and repair, including planning principles for sustainability

These three major recycling possibilities aren't in fact equal, but rather a ranking exists among them, with reconstruction at the highest level, then restoration, and finally repair. Reconstruction is a practice of upgrading a used commodity to a minimum to the new product quality requirements and providing the resulting product a warranty that would be at least on par with that of the new produced alternative [29]. At present, reconstructing is typically profitable for large and complex mechanical and electromechanical products that follow an extremely stable technological process and for materials and components that are expensive to manufacture or can become expensive in the future. The value of component reuse in relation to dismantling costs makes manually dismantling these products worthwhile, allowing the reconstruction of profitable products. Reconstruction can be differentiated from repair and renovation, at four key points:

- Reconstructed products have a guarantee equal to that of new alternatives, while repaired and renovated have lower guarantees. Typically, with the renovation, the warranty applies to all important parts that are damaged, and the repair is only applicable to the repaired part.

- Reconstruction requires more effort than the other two methods, with the outcome that the standard and efficiency of its commodities appear to be greater.

- Reproduced devices lose their identities when rebuilt and refurbished maintain their own as all parts of a commodity are recycled in the reconstruction and new ones are replaced by people who can no longer at minimum return to their initial standards.

- Reconstructing may include upgrading the product used beyond its original specifications, which is not the case for repair and renovation.

The main advantage of reconstruction is that it allows to combine the low price and good quality of the products, especially when the reconstruction also involves increasing the performance and quality of the products used beyond their original standards when these were new. Xerox is a typical example of successful reconstruction since its copiers are typically subjected to seven life cycles. This means that seven revenue streams are generated from the construction of a single product, and materials are diverted from landfill or recycling at least six times [30]. The disadvantage of reconstruction with regard to smaller product recovery processes is its higher cost because of the more resources used and most of the work required to get the reconstructed product.

Thus, there are many products where the reconstruction would have a prohibitive cost given the current reconstruction technology and the basic knowledge required. Major home appliance manufacturers, such as Lec Refrigeration and Merloni, implicitly suggest that rebuilding of household appliances has a prohibitive cost, at least within EU. The main reason is cost of manual labor involved in reconstructing and additional costs, such as costs for testing with safety standards that are accurate. Cost of these tests in new production can be limited by "run" per batch, but in reconstruction, tests have to be done individually.

3.1 Planning principles for sustainability

Several basic rules of sustainability management can be enforced to all types of commodities, including electrical and electronic equipment. These can be illustrated as follows:

- A real issue, originating from an actual problem and working on solutions that are culturally, ecologically and financially advantageous
- Research, specifically, of processes and facilities, not commodities, such as not beginning with the development of a washer and dryer, with the air to satisfy consumers but finding environmentally friendly ways of cleaning clothes
- Participation of users, operators and various experts in design process as much as possible
- Attempting to investigate all the different dimensions and criteria, as well as prioritizing hierarchies according to the timing and scope of the project

3.2 Life cycle consideration

Considering that the lifespan is an integral and absolutely essential aspect of DFS, this covers all aspects of the lifespan of a product, from resource extraction to the phases of use and end of life. In this sense, policies for reuse and recycling including biodegradation, incineration and final disposal are being introduced. DFS seeks to continually improve the sustainability of the entire system at all levels of the lifespan of the commodity, such as removing harmful chemicals, improving efficiency and performance, promoting reuse and recycling. In addition, DFS analyzes the context of use and the systems in which products and services are added to the life cycle with the prospect of a sustainable product-service-system design [31]. When designing sustainable product service system, the whole mentioned system will be considered.

3.3 EEE usage phase

The minimization of energy during the use of electronic products is the main objective of the EUP Directive. The EUP criteria for the different product categories cover for example the energy consumption during the use of the appliances and the standby power consumption. Although energy consumption thresholds are mandatory for all products covered by the Directive, no design strategies or additional measures to encourage the so-called sustainable user behavior are defined.

Therefore, the use process consists of several phases, such as commodity acquisition, start-up, utilization, maintenance restoration and recycling, and consumer attitudes at any stage that is hard to anticipate. During the engineering process, extra effort should be undertaken to influence customer behavior, for example, to provide characteristics and data through the packaging in the commodity and customer information to cultivate customer's economically viable attitude.

Design strategies for optimizing the use phase include the following:

- Knowledge delivery, support infrastructure and possibilities for environmentally responsible use and recovery after use

- Maintenance operations architecture

- Incorporation of functionalities to reduce power expenditure (e.g., the display of vacuum cleaner effectiveness data—indication signal in case a filter change is required)

- Inspiring and persuading buyers to reconsider their practices, such as not using dryers and drying clothes outside, laundering only under maximum load and at lower temperatures, etc., and additionally marketing them in an enjoyable and pleasurable manner, or issuing customer and group rewards [32].

4. Repair of products in industry; tools and rules for sustainable design

4.1 Repair of products in industry

One of the most typical product repair applications, which will be discussed in this section, is that of computers. The repair of computers and other office products, such as printers, has been going on for over 20 years and is not guided by the original manufacturers but by independent experts who have identified a commercial opportunity. Most manufacturers do not yet have the repair as a priority for serving their customers or the market, so demand is still met mainly by independent providers. Reopenable computers and print products can generally be categorized into three categories: repaired, renovated and reconstructed. There are currently a growing number of manufacturers who have put in place procedures to provide second-hand equipment to their customers, with some using "home-based" services and others involving independent experts as service providers. In the absence of legislation and standards, acceptable practices will be different among all suppliers of second-hand equipment (whether manufacturers or specialists), but the following descriptions provide a guide to product expectations within the three categories and throughout the computer market.

4.2 Tools and rules for sustainable design

To facilitate sustainability planners, there is already a wealth of tools and methods available to help integrate the environmental, social and economic aspects of planning processes. The most complex tool is Life Cycle Assessment (LCA), and the simplest are the rules of thumb that experts have formulated to give guidance to the design process. Through cumulative knowledge acquired for decades, various thumb rules have been developed concerning the following:

- Longer-lived products that consume significant amounts of energy, fuel, water, and other consumables that, during their lifetimes, very often, have significant environmental impacts occurring during use phase (making the reduction in consumption during use phase)

- Longer-lived moving products, for example, energy-consuming vehicles in practice, where the weight and other effects on energy consumption in the active/in-use phase are usually more important and consumables, for example, products with a very short life span that are lost or dissolved during use, where it is important to be non-toxic and biodegradable

5. Reusability and after that

5.1 Repair

Any fault or damage intervention, identification and correction is called repair. A mechanical or electrical repair will return a commodity to a working state, while a decorative repair can inflict cosmetic damage to the outside appearance and/or spots (e.g., crack, stain, scratch or rupture). A certified technician or a service center could restore a commodity regionally of the manufacturer or specialist. The test is only performed to ensure that the repair has eliminated or corrected the particular defect identified. Repairs are inherent activities in more extensive repair or rebuild processes.

5.2 Renovation

Renovation is one of the two processes associated with most reused products. It is carried out in a factory with functional specifications, involving a larger set of tools, cleaning solutions, solvents, paints and other surface treatment options. The upgrade is described as the return of a used product to a performance or quality greater than it was when it was brand new. While the refurbishment process does not seek to increase the original manufacturing capacity of the product, larger spare parts may be added if genuine spare parts are no longer available, or later higher-capacity components are of comparable cost. A refurbished product generally has a limited warranty, depending on the supplier (original manufacturer or independent specialist).

5.3 Reconstruction

Reconstruction is more complex than refurbishment and is a thorough, complete dismantling and reassembly process that returns a used product at least to the originally determined state. Depending on the processes of the remanufacturer (either original manufacturer or independent specialist), the disassembly process can either preserve the identity of the original product (through serial number) or provide a completely new identification system (supported by a new serial number).

Reconstruction includes detailed cleaning, testing and diagnosis of all dismantled parts. Components, depending on their commercial viability, are either repaired or replaced. Repairs to components or sub-assemblies are performed by the remanufacturer or shipped to a specialist on the product. Upgrades are also provided for parts of the hardware when commercially available, and any other changes to the software or logistics infrastructure should also be included in the rebuilding process.

Restoration is conducted at a factory location with appropriate sets of equipment and measurements similar to that used in initial production with guidance found in the method of audit documentation. Due to the complete removal of the items, the factory default configuration must be deleted or modified. Furthermore, new additions and improvements can be introduced in order to provide commodities with the latest available innovations compared to standard prototypes. Refurbished commodities can, therefore, be identical to current production versions, checked at the same standard, and commonly sold as "new" with a complete or revised insurance.

It has to be mentioned that the standard engineering term of reconstruction in the field of information technology is more like rebuilding, as very few are actually reconstructed in the same way a component was originally constructed. Most computer and printer vendors will use specialized manufacturers to build key components and subsystems (such as processors, memories, optical drives, and hard drives), and the cost of replacing with a modern component is generally lower than repairing older defective product.

5.4 Upgrading

Repair can be part of the renovation or rebuild process. Upgrades can be developed to meet customer-related issues, or programmed events in the product life cycle, especially when the product is complex and designed for long service life. Improvement tends to increase or boosts the efficiency of the device by upgrading its capabilities or performance, along with changing or incorporating components or modules to enhance the capabilities of the original model. Just like the repair, the examination is minimal and then only to ensure proper implementation and operation of the improvement.

Several adjustments may improve the item's capability further than the level of initial manufacturing processes, and some may even push a commodity to the latest model standards. It is dependent on an item's evolutionary interoperability that relies on the prototype technical adaptability and the ability to improve a model throughout its lifespan.

Improvements could also emerge from the absence of the initial product, which may result in improvements caused by a lack of option in repair and refurbishment.

5.5 The secondary product market

While the reconstruction of electrical equipment cannot mitigate ecological advantages and competitiveness, it can be accepted regarding social aspects such as addressing inequality, unemployment and absence of qualifications. The policy choice to be decided throughout the context of second-hand trade functions for some types of commodities such as home equipment should be whether or not ecological issues and their decreased productivity can be overcome by their tremendous social gains and re-operating ecological gains. Therefore, the ecological risks are likely to overshadow the significant social benefits. Some socioeconomic positive effects of resale business operations comprise employment advancement, improving the overall professional career for the regional society and individuals reselling second-hand consumer commodities, supplying basic goods to disadvantaged people who would otherwise be unable to access them, and providing with employment access to unqualified individuals. The socioeconomic consequences of resale business procedures may be defined via the efforts of several associations working with homeless people and several marginalized communities (https://emmaus.org.uk/) [33]. Organizations get donated products that are re-running and help the homeless to reopen them under their supervision. Their key advantages include the following:

- The homeless benefit from having a roof above their heads, paid employment, self-confidence and new skills that will help them start over.
- Organizations continue their various charitable goals.
- Jobs are created for the technical staff that oversees former homeless people.
- Poor people benefit because they can afford the purchase of goods.
- Jobs are being created.

5.6 Variability of standards and quality of renovation and reuse

Despite the fact that there is EU legislation for processing commodities that have reached the end of the lifespan stage, EU legislation does not exist today. Neither are other wealthy nations or areas motivated toward refurbishment or reconstruction to recycle IT commodities. Existing WEEE regulations include obligations and standards for optimal material recycling after it has been marked as scrap, but still there is very little to direct consumers as well as producers on reuse or prolonged utilization.

Without the regulations, the system of manufacturing requirements is very minimal and there are variations in the rate of re-operation and production quality with almost all of the construction in the control of autonomous professionals. Autonomous dealers, motivated by market opportunities, may very well typically try relatively cost-effective options to bring a device to function because then they can take advantage of a secondary efficient lifespan, resulting in competitive and dynamic business availability. Certain industry associations representing manufacturers and independent suppliers have tried to clarify the process, but are currently not developed in recognized national or international standards that can be independently controlled to provide recognized levels of accreditation.

5.7 Qualitative criteria for reuse and certification of reuse centers

As previously stated, in the sector of discarded IT appliances, there is little legislation, so specifications and standard requirements differ for all vendors on the sector of old IT hardware. The simpler present legal system applies to the purchase of goods, acknowledging that a commodity should not be altered and needs to be in line with the specific intent and characterization. Most of the items sold in stores are therefore defined as being partly "used," yet without more explanation. By identifying their commodities as former-approved, certain vendors would only further distinguish their offers formerly-declared, formerly-borrowed, etc. This is generally the case for the newest used equipment below 12 months. Occasionally, some manufacturers sell excess product or product inventory through their used product channels and at lower prices, even if they are not open or new.

Some suppliers and broader individual distributors can mark their commodity rebuild procedures with some other industry certifications, such as international standards ISO, CEN, or national standards such as BSI or DIN. In particular, in the UK, BSI offers a protocol that includes descriptions and guidelines for the repair and future sales of used IT appliances [34]. This template has the acronym MADE (made for assembly, disassembly and end of life). This includes descriptions of procedures for re-assembly levels and re-launch of the equipment back to the market. At present, this standard serves as a voluntary industry guide without any certification or accreditation procedure that will confirm the correct practice by the supplier (be it manufacturer or independent).

Many of the suppliers of used equipment also have a waste permit for their reprocessing facilities to ensure compliance with legislation and to properly dispose of the waste generated in the re-operational processes. As with some repair work, some suppliers of used equipment may outsource the recycling operations to third parties.

6. Design issues in reconstruction and for reuse and recycling of EEE

6.1 Design issues in reconstruction

Improvement of refinancing and refurbishment would require changes in the architecture of the product as the architecture is the lifespan characteristic of the commodity that has the greatest influence on ecological burden [35] and also defines the commodity's capacity. That is, it will immediately raise the particular product cost and it will initially be costly but it will lead to long-term sustainability, bearing in mind the relative cost of waste management and many other environmental requirements. One major problem at this point is the lack of know-how of designers for designing products for reuse. A key issue in product design for reuse is to avoid features that prevent the product or component from being returned at least in its original state of operation. These include the following:

- Non-lasting component that may result in rupture during reconstruction (construction, maintenance or upgrade) as well as a limitation during use to the degree that the commodity is inadequate for "refurbishment"

- Involvement of techniques that prohibit element isolation or are likely to result in element destruction during separation: epoxy resin welding, for example, can be used to promote fast assembly but will prohibit disassembly without harm, leading to an increase of even further innovation in recycling or reuse

- Characteristics involving banned elements or techniques of storage or anything that might constitute the process costs unviable economically

Several of these main refurbishments and reuse considerations, though, bypass the influence of the manufacturer. The most critical of these are the rules, demands and restriction procedures of factories. Laws and regulations will have a significant impact as they allow businesses to raise the value added of their commodities and increase the cost of disposal. This can also motivate companies to produce refurbished commodities. Furthermore, if laws prevent the use of a chemical, the products that contain it cannot be re-imported into the marketplace and thus will not be reused. Refurbishment and reuse are only acceptable if the revived item has a demand. Fashion-affected goods are improper since consumers may choose the latest offering irrespective of the refurbished's price and quality. Many consumers demand modern goods as fashion options, so goods are usually less attractive in terms of renovation and reuse, particularly those that require a fairly low preliminary cost or are in prestigious positions in residences. Manufacturers' prohibitive practices, such as patents, property rights and anti-competitive processing, also prevent refurbishment and reuse. For example, some printer manufacturers have designed the ink cartridges to self-destruct when they are empty, thus preventing their rebuilding. However, if the old products do not exist to get rid of them or take good aspects of existing second-hand products, then the technology to produce new parts becomes obsolete, and therefore refurbishing the product will be impossible.

6.2 Design for reuse and recycling of EEE

According to the EU waste hierarchy, the mandatory priorities for EEE are prevention, reuse, recycling and other forms of recovery and finally disposal, if economically and environmentally feasible.

WEEE regulations also contain a strict requirement to promote consumer reuse and recycling pursuant to Section 4: Product Design. Member states are obliged to take action to stop manufacturers from the use of particular design characteristics or particular production processes unless such characteristics or processes are legally required or the actual advantages are immensely beneficial. In product design, to consider whether any of these approaches is most effective, the options for reuse of an item should be analyzed and correlated to raw recycling and dismantling policies.

The concept of reuse may include either individual components or the whole of the product, depending on its age and condition. Reuse can take place for the same purpose in the same system or to serve another purpose. During the design of the concepts or possibilities for reuse, the following criteria must be taken into account [36]:

- Technical criteria
- Quantitative criteria
- Economic criteria
- Time criteria
- Innovative criteria
- Delivery criteria
- *Criteria* for the compatibility of reusable devices with the standards of new EEE
- Other criteria, for example, market behavior, obligations, patents and property rights. Design strategies that support reuse and recycling include:
- Structured design and standardization of components
- Longevity design
- Design for recycling. Selection of recyclable materials, low material diversity, low toxicity, marking of materials and ease of dismantling:
- Recovery of valuable materials in electronic products

- Disassembly of components containing dangerous substances
- Disassembling components that obstruct recycling technologies upgrade planning. It includes technical upgrade
- Design dismantling and assembly
- Design concepts that make wear and tear of parts detectable and visible.
- Provision of instructions and information on recyclers and disposal instructions for end users
- Design of product-service systems. Maintenance, recovery and repair, lease upgrading, leasing, exchange, centralized services

All these strategies have as a common feature that they are very dependent on the systems around them. Waste, waste management and logistics systems affect recovery and reuse rates, while end-of-life design requires real planning elements and communication with the end-of-life industry.

Recycling is usually the ecological goal as it preserves the assets (materials and energy) expended in the commodity throughout production. This is accompanied by the recovery cycle recycling of parts and components. Such techniques include dismantling that is non-destructive. The most standard procedure at the moment, however, is product recycling whereby resources are acquired and structural specifics are destroyed. This technique makes catastrophic modification feasible and is a standard procedure for retrieving precious materials, for example, platinum and gold.

The life cycle reuse and recycling methods of a material can be applied as follows:

- Improve consumer recycling alternatives, namely recycling and end-of-life management, item recycle and recycling solutions (prolonged service life via simple and rapid substitute of broken parts) and recyclable pieces; consumable parts and materials throughout usage cycle.
- Improve after-life recycling involving recovery/exchange programs, second-hand sales, non-destructive dismantlement and refurbishment recycling of the item and minimizing the components used in the commodity.

Financial factors are important in the design and analysis of reuse and recycling approaches, the market price of recycling versus the price of new products. Luckily, increasing resource international market costs advantage reuse and recycling. However, we must also be interested in the effects of reuse, for example, when much better effective and eco-friendly new technologies and services become accessible, the replacement of obsolete goods might be environmentally less harmful.

7. Effects on renovation and reuse

Traditionally, safety, performance and cost were the key factors in the decision to build a product. However, globally changing business conditions force organizations to re-analyze their strategic decisions. Thus, additional factors such as raw material costs and environmental legislation are taken into account in planning and construction decisions. This leads to a shift in factors that affect reuse and refurbishment. Two key factors are the shift from the sale of the product to the sale of capacity (shifting to 'product-service' systems [37]) and the shifting of some companies away from production to the assembly or redemption of segments. As for the first, traditionally, manufacturers sell the products to their customers, so that there is a transfer of ownership from the manufacturer to the customer. Today, some manufacturers choose to retain the ownership of their product and instead sell the product's capability to the customer. One such example is the "provision with time" in the aerospace industry. The manufacturer acts as a service provider and assumes any risks associated with the failure of the product. As the consumer buys only the capacity guarantee, the interest is focused on customer satisfaction with the provided capacity, and so the product age (number of life cycles) becomes less important. Renovation and reuse reduce the costs of organizations that adopt the business model of service, for example, maintenance costs are reduced through the use of refurbished and reused components, and whole remanufactured or refurbished machines can be used instead of more expensive new ones. In the latter case, some producers, in order to reduce costs, buy components from countries with lower labor costs and simply assemble these parts. This however, leads to the loss of the required engineering skills for the reconstruction.

8. Availability of information on components, materials and methods of repairing products (including influence of scarce resources on sustainable EEE design)

There is a clear difference between the position of the original manufacturer and the independent specialist. The original manufacturer will have access to all original manufacturing information as well as subsequent mechanical changes throughout the product's production path. Most manufacturers provide a dedicated production line to keep their interest in producing new products, but some companies such as Ricoh are running re-operational products on the same lines as new products.

The required detailed info usually provided by the company would assign the opening to an associated operator who can work on-site at the location of the producer or at one's own venue. The producer could also have links to authentic products of spare parts and distributors, and perhaps provide new and existing parts to update the re-launched devices.

Autonomous repair professionals are far more constrained to restart because they work without the initial producer license. They have no links to method or item suppliers' information and therefore need to gain substantial technological capabilities in other manners. Necessary parts are bought from the open market, whether new or even used, and often specifically from licensed and autonomous service suppliers, or from the distribution channel of the supplier or its associate. In certain situations, full systems for parts acquisition should be acquired to allow components to be replaced for products to be re-operated. Product knowledge and experience may be acquired through the recruitment of staff previously employed by the original manufacturer or their authorized sales and repair partners.

Depending on the size of independent experts, resumption capabilities vary with the size and depth of the process, but even larger independent companies cannot invest in full reproduction of the original maker's production or restart environment. Repair and restoration methods will generally be similar between the manufacturer, the authorized representative or the independent specialist, to test the product, re-operate at the required level and prepare for use. Independent experts may generally have the most efficient line to bring a used product back into a repaired or refurbished mode, while the manufacturer may choose to invest more in time and cost of re-operation to provide a good quality of used product with a similar warranty.

All participants are responsible through decision-making phases, which assess the product's re-launch at different stages to guarantee an environmentally friendly degree of re-operation is selected that facilitates second-hand market at a gain rather than a loss. Many producers will not pursue the high secondary market revenue of used appliances since they are willing to give independent companies the first and most favorable deal to provide as much customer service as feasibly possible.

8.1 Selection of sustainable materials and processes

The choice of materials and processes is another important element of DFS. There are eight key criteria for these purposes: (1) consumption of resources, (2) energy consumption, (3) dangerous substances emissions, (4) origin and transport, (5) aspects of life span, (6) waste generation, (7) biodiversity and protection of natural areas, and (8) social aspects.

9. Conclusions

Currently, in the sector of recycling of waste electrical and electronic equipment, there is a growing need for rigorous and comprehensive policies to reduce the ecological effects of WEEE. Challenges and opportunities involve ecological pollution, mineral resources scarcity, waste treatment, and landfill deterioration. Thus, WEEE's regeneration and reconstruction are rather essential owing to its huge environmental consequences. Sustainability is not a stationary condition, but a system of dynamic equilibrium between the human environment and the ecosystem. Cofactors that play an influential role is sustainability strategy and environmentally friendly design techniques, dismantlement and reuse architecture, and what they indicate for home appliances. Policies and regulations as well as standards should be identified in the light of opportunities, techniques and equipment for environmentally friendly architecture of consumer electronics, with specific focus on reuse, recycling, choice of environmentally friendly new materials, and finally limited resources.

Author details

Panagiotis Sinioros[1], Abas Amir Haidari[2], Nikolaos Manousakis[1], Michael Lasithiotakis[3]* and Ourania Tzoraki[2]

1 Department of Electrical Engineering, Technological Educational Institute of Piraeus (TEIPIR), Greece

2 Marine Science Department, University of the Aegean, University Hill, Mytilene, Greece

3 Environmental Radioactivity Monitoring Department, Greek Atomic Energy Commission (EEAE), Athens, Greece

*Address all correspondence to: michalis.lasithiotakis@eeae.gr

References

[1] Ijomah WL, Chiodo JD. Application of active disassembly to extend profitable remanufacturing in small electrical and electronic products. International Journal of Sustainable Engineering. 2010;**3**:246-257

[2] Hawken P. The Ecology of Commerce—A Declaration of Sustainability. New York: Harper Collins; 1993. pp. 45-46

[3] Council Directive 91/689/EEC of 12 December 1991 on hazardous waste. Document 31991L0689. 91/689/EEC - EUR-Lex - Europa EU

[4] Council Directive 94/31/EC of 27 June 1994 amending Directive 91/689/ EEC on hazardous waste. Document 31994L0031. EUR-Lex - 31994L0031 - EN - EUR-Lex - europa.eu

[5] Stern N. The Economics of Climate Change. The Stern Review. Cambridge, UK: Cambridge University Press; 2007

[6] European Environment Agency. Environmental Pressures from European Consumption and Production. 2007. EE A Publication TH-78- 07-137-EN-D

[7] World Commission on Environment and Development. Our Common Future. Brundtland Commission. Oxford, UK: Oxford University Press; 1987

[8] United Nations, Agenda 21: The Earth Summit Strategy to Save Our Planet. 1992. Document E. 92-38352

[9] Papanek V. Design for the Real World: Human Ecology and Social Change. New York: Pantheon Books; 1971

[10] Papanek V. The Green Imperative: Natural Design for the Real World. New York: Thames and Hudson; 1995

[11] Tischner U, Hora M. Chapter 17 - Sustainable electronic product design. Waste Electrical and Electronic Equipment (WEEE) Handbook. 2nd ed. Woodhead Publishing Series in Electronic and Optical Materials. 2019. DOI: 10.1016/ B978-0-08-102158-3.00017-3

[12] Charter M, Tischner U. Sustainable Solutions: Developing Products and Services for the Future. Sheffield, UK: Greenleaf Publishing; 2001

[13] NRC. Buy Recycled Guidebook. 1999. Available from: http://www.nrc-recycle.org/brba/Buy_Recycled_Guidebook.pdf [Accessed: 21 November 2003]

[14] Ijomah WL. The application of remanufacturing in sustainable manufacture. Proceedings of ICE–Waste and Resource Management. 2010;**163**:157-163

[15] Chick A, Micklethwaite P. Obstacles to UK Architects and Designers Specifying Recycled Products and Materials, Design History Society Conference. Aberystwyth: The University of Wales; 2002

[16] Ijomah WL, Childe SJ. A model of the operations concerned in re-manufacture. International Journal of Production Research. 2007;**45**:5857-5880

[17] Hora M. 'Consumer Advisory Services' and 'European Directive on Waste from Electrical and Electronic Equipment (WEEE)' in 'Policy Instruments for Resource Efficiency–Towards Sustainable Consumption and Production'. Eschborn: GTZ (Deutsche Gesellschaft für technische Zusammenarbeit), UNEP/Wuppertal Institute CSC P, Wuppertal Institute; 2006

[18] European Commission. Directive 2005/32/EC on the Eco-design of Energyusing Products (EuP): Environmentally-friendly Design of Energy-using Products: Framework

Directive for Setting Eco-design Requirements for Energy-using Products [Online] 2005. Available from: http://ec.europa.eu/enterprise/eco_design/index_en.htm

[19] European Commission. Manufacturing Visions Report No. 3, Integrating Diverse Perspectives into Pan-European Foresight, Delphi Interpretation Report, Contract No. NMP2-CT-2003-507139-MANVIS, 11/2005 [Online]. 2005. Available from: http://forera.jrc.ec.europa.eu/documents/Final_Report_final.pdf

[20] Communication from the Commission to the European Parliament, the Council, the European Economic and Social Committee and the Committee of the Regions Public procurement for a better environment {SEC(2008) 2124} {SEC(2008) 2125} {SEC(2008) 2126}/* COM/2008/0400 final */. Document 52008DC0400. Eur-lex - :52008DC0400 - EN - Eur _Lex

[21] Monti M. A New Strategy for the Single Market at the Service of Europe's Economy and Society. Report to the President of the European Commission José Manuel Barroso [Online]. 2010. Available from: http://ec.europa.eu/bepa/pdf/monti_report_final_10_05_2010_en.pdf [Accessed: 30 April 2011]

[22] [Online]. Available from: http://www.ecolabelindex.com/ecolabels/?st=category=electronics [Accessed: 25 June 2011]

[23] US Government. Dodd–Frank Wall Street Reform and Consumer Protection Act ('Dodd-Frank Act') [Online]. 2010. Available from: http://www.sec.gov/about/laws/wallstreetreform-cpa.pdf, http://www.sec.gov/spotlight/dodd-frank/speccorpdisclosure.shtml [Accessed: 25 June 2011]

[24] ISO/TR 14062. Environmental Management—Integrating Environmental Aspects into Product Design and Development. International Organisation for Standardization [Online]. 2002. Available from: http://www.iso.org/iso/catalogue_detail?csnumber=33020 [Accessed: 24 July 2011]

[25] ISO 14006. Environmental Management Systems—Guidelines for Incorporating Ecodesign. International Organisation for Standardization [Online]. 2011. Available from: http://www.iso.org/iso/iso_catalogue/catalogue_tc/catalogue_detail.htm?csnumber=43241 [Accessed: 24 July 2011]

[26] KIA. Environmental Certificates [Online]. Available from: http://www.kiaglobal.com/sg/en/NewsPromotions/NewsRelease/013303.html [Accessed: 24 July 2011]

[27] Daimler. Mercedes Benz Models with Environmental Certificates [Online]. Available from: www.daimler.com, http://www.daimler.com/dccom/0-5-1312394-1-1312442-1-0-0-0-0-0-36-7145-0-0-0-0-0-0-0.html [Accessed: 24 July 2011]

[28] Sinus. The Sinus Milieus in Germany [Online]. 2011. Available from: http://www.sinus-institut.de/en/ [Accessed: 30 April 2011]

[29] Ijomah WL. A model-based definition of the generic remanufacturing business process. [PhD dissertation]. UK: University of Plymouth; 2002

[30] Gray C, Charter M. Remanufacturing and Product Design [Online]. 2006. Available from: http://www.cfsd.org.uk/Remanufacturing%20and%20Product%20Design.pdf [Accessed: 7 November 2011]

[31] Tukker A, Tischner U. New business for old europe, product-service development. In: Competitiveness and Sustainability. Sheffield, UK: Greenleaf Publishing; 2006

[32] Agarwal S, Nath A. Green computing - A new horizon of energy

efficiency and electronic waste minimization: A global perspective. In: International Conference on Communication Systems and Network Technologies. Katra, Jammu; 2011. pp. 688-693. DOI: 10.1109/CSNT.2011.148

[33] Cole C, Gnanapragasam A, Cooper T. Towards a circular economy: exploring routes to reuse for discarded electrical and electronic equipment. Procedia CIRP. 2017;**61**:155-160. ISSN: 2212-8271. DOI: 10.1016/j.procir.2016.11.234

[34] BSI, BS 8887-2:2009—Terms and definitions, BS 8887-1:2006—General concepts, process and requirements. Produced by British Standards institute technical product specification committee (TDW/004/0-/05 Design for MADE BSI). 2009

[35] Graedel TE, Allenby BR, Comrie VR. Matrix approaches to abridged life cycle assessment. Environmental Science & Technology. 1995;**29**(3):134-139. Available from: https://pubs.acs.org/doi/pdf/10.1021/es00003a751

[36] Rosemann B, Brüning R, Enderle B, Schmidt K, Spengler TS, Plumeyer M. The VDI 2343 guideline gives recommendations for the concerned parties: Part reuse. In: ISWA World Congress, Hamburg, Deutschland; 2010. Available from: https://eref.uni-bayreuth.de/id/eprint/32026 [Accessed from: 15-18 November 2010]

[37] Sundin E, Lindahl M, Ijomah W. Product design for product/service systems: Design experiences from Swedish industry. Journal of Manufacturing Technology Management. 2009;**20**:723-753

Chapter 6

Adaptation: A Lens for Viewing Technology Transfer in Construction Site Management

Aghaegbuna Obinna U. Ozumba and Winston Shakantu

Abstract

This book chapter presents the results from a series of studies which explored the use of the technology transfer (TT) subconcept of adaptation to explain the uptake of recent information and communication technologies (ICT) in the construction industry. The specific focus is the management of construction site processes. The studies explored the need for management process enhancement, availability of relevant information and communication technologies, occurrence of such technologies in construction site management (CSM), influencing factors, and challenges to their adoption in construction site management. Results from stages in the phased methodology are used to generate certain hypotheses that are based on analysis of primary and secondary data. Insights from testing the hypotheses and findings from the series of studies are used to model an adaptation-based understanding of the transfer of information and communication technologies in construction site management. While using site management as the specific focus, the study contributes an understanding that is relevant to the construction industry and other project-related environments.

Keywords: adaptation, adoption, construction, management, ICT, IT, site management, technology transfer

1. Introduction

The focus of this chapter is the use of the concept of adaptation [1], which is a subconcept of technology transfer (TT) [2, 3], to explain the uptake of recent digital and automation technologies, here classified as information and communication technologies (ICT), in the built environment. The chapter looks at management functions, specifically focusing on construction site management (CSM) and its process. The chapter is based on a series of studies within a doctoral research project. The following themes were explored: the construction industry in a continuously changing environment, implications of changes and increasing complexities of construction sites, challenges and strains to site management, human weakness and limitations in managing construction sites, and technology in an evolving construction industry. It was possible to highlight the realities being faced by the building and construction sectors, and the use of ICT to enhance the site management process, in the face of attendant challenges and limitations.

The construction sector is arguably a big role player in every economy, and the CSM process is strategic in the construction process, being the management of physical delivery of the construction product. The term product refers to any type of physical facility, structure, and infrastructure. Despite appreciable research, technology uptake in the construction industry has been regarded as slower than other sectors. While there is increasing uptake of ICT in construction, there is a persistent shortfall in the exploitation of benefits offered in recent outputs of such technologies. This is especially true in developing countries such as South Africa, which was used as context for the research project, where studies such as [4, 5] suggest the need for more ICT uptake. Without an increase in utilisation of ICT, potential benefits for various aspects of construction, including CSM, remain untapped. The situation is undesirable, considering the sector's need for performance in the face of persistent demand for more development. The construction sector is commonly used by governments, as a vehicle for physical and social development, thereby increasing the limitations and pressures experienced in managing projects. The implications for CSM, the need for management process enhancement, and the relevance and potential of recent ICT and its slow uptake in construction formed the research background. For scope, three focus areas were chosen: site materials management (SMM), site health and safety management (H&S), and site information management and communication (IM&C). The research problem was articulated as follows: Exploitation of the adaptability of recent ICT developments to improve the site management process is suboptimal, resulting in lost opportunities that manifest as lapses in information management and communication, H&S, and materials management.

Thus, the broad research aim was to investigate the exploitation of the adaptability of recent ICT developments to improve CSM and to build an understanding of the nature of this exploitation through the lens of adaptation, in order to ultimately generate propositions for addressing the suboptimality.

1.1 Initial theoretical studies in the research project

The research project included an extensive literature review and a multistage investigation strategy, set in South Africa, spanning an initial time frame from 2008 to 2013. From 2014 to 2019, more research was carried out, generating papers and articles, extending the literature, and addressing the many sub-objectives of the wider research project. Between 2008 and 2019, the research project resulted in 14 research papers, of which 2 were based on extensions of the research, by supervised students, 4 (4) journal articles, and a doctoral thesis. A historical perspective of publications from the initial studies is presented in **Table 1**.

Except for publications in 2010, all papers in **Table 1** are theoretical. This period saw an extensive theoretical exploration of various areas deemed relevant to the research, using secondary data and limited primary data. Results of [6–8] substantiated the need for an improved CSM and the need to use ICT for that purpose. This was followed by substantiation of CSM needs in relevant aspects such as SMM and IM&C [10, 11]. The need for user-centredness, in terms of usability, acceptability, ICT readiness, and/or skills issues, was substantiated in [9, 12]. In 2010, the relevance of contextual and compatibility issues such as the developing country context, local ICT education, and indigenous technology cultures was addressed, with empirical data in [13, 15]. From the studies in **Table 1**, it was determined that in the face of unparalleled pace of output from the ICT sector, it was reasonable to query the noted slow pace of ICT adoption in construction. In order to explain the situation within the scenario of CSM, various useful subconcepts of innovation [16] and technology transfer were considered, as discussed in section 2.

Reference	Publication	Year
[6]	Improving site management process through ICT	2008
[7]	Achieving ubiquity in the site management process: a theoretical study of the potential for innovative ICT solutions	2008
[8]	Improving materials management through utilisation of information and communication technology	2008
[9]	Balancing site information and communication technology systems with available ICT skills	2009
[10]	Information and communication technology – based application of 'just-in-time' (JIT) to internal logistics on site	2009
[11]	Enhancing on-site communications by adaptation of multimedia systems: looking into the future of construction	2009
[12]	Improving people-centeredness in H&S risk management through ICT	2009
[13]	Cutting-edge technology for construction ICT in a developing country	2010
[14]	Information and communication Technology education within South African built environment schools	2010
[15]	Indigenous iron technology evolution: lessons from Uzu culture of Awka in Nigeria	2010

Table 1.
Historical perspective of initial publications from the research project.

2. Summary of theoretical basis for the research

In the research, information was accessed from historical and more contemporary literature. An appreciable effort was made to understand the relevant and key concepts and theories, including theory of reasoned action (TRA) and theory of planned behaviour (TPB) [12, 13]; technology acceptance model (TAM) [14, 15]; unified theory of acceptance and use of technology (UTAUT) [12]; technology-organisation-environment (TOE) framework [16]; MIT90 model of organisational information system innovation [17]; business process adaptation models, referred to as business process automation (BPA) [18, 19]; structural equivalence theory [20]; threshold innovation theory [21]; and the process theory, among others [22]. Considering the research focus, technology (i.e. ICT) and its transfer were determined as key issues. The research inherently centred around the parent concept of technology transfer. Hence, relevant theories and concepts were explored, including technology and technology culture [23, 24]; TT [2, 3, 25]; innovation [11, 26]; diffusion [27, 28]; diffusion of innovations [27, 29, 30]; adoption theory [29, 31]; adoption and user acceptance [29, 31]; the function of knowledge in adoption and innovation diffusion [29]; and adoption model based on the contagion concept, social influence, and social learning [32]. Going through the plethora of relevant concepts and theories, none adequately or fully explained the interaction of people and ICT in the unique area of CSM. Further deliberation led to the concept of adaptation [33, 34], which was employed as the main perspective for understanding the results.

2.1 The adaptation concept

The word adaptation appeared as a borrowed term, originating from Latin in the thirteenth century. The term was used in both the tangible and the abstract sense. The modern use of the term began in the sixteenth century [33]. It is derived from a Latin root *apt* or *apt-us*, which refers to something that is appropriate, or suited for purpose, or the context, which is in turn derived from the root *ap-ere*, meaning to fasten, affix, clip onto, or attach. It can be defined as the process or the product of these actions [34]. Adaptation is also described as making fit, and it will involve more than simply imposing something into a different setting [35]. The definitions allude to adaptation being purpose-driven [34]. Adaptation as a generic concept has expressions in disciplines such as psychology, biology, anthropology, sociology, and geography [33]. By extension it has been used to explain certain patterns of TT. While many authors allude to the concept in defining and describing TT, there are others such as Brooks in [25] '...explicitly theorizing adaptation as a sub-concept of...' TT, being a process of moving technology, '...from more basic scientific knowledge into technology or adaptation of an existing technology to a new use'. Furthermore, in [36] TT is described as having two possible routes: by means of imitation and through an analogy, which means that '...the technology must be adapted before it can be adopted...someone must see an analogy between the characteristics of the original invention or innovation and the requirements of the new situation'. The outline of the TT process in [25] shows that the receiver of technology will either repair and maintain; modify and adapt; or design and produce new equipment or products, based on the technology. Adaptation is also described as a fundamental concept in technology transfer in [37]. Here it highlights the significant movement of technology between different sets of users and the movement of technology for the purpose of application towards addressing another problem apart from that for which it is originally designed. Furthermore, it is arguable that the process of diffusion in some cases may produce a profoundly different result from the original technology. There could be a case of slight modifications which may be a 'fine-tuning' requirement. The technique of applying the technology could also be addressed differently in varying degrees.

With TT as the parent concept, and the subconcept of adaptation as the lens for the research, key assumptions held in the research include the following: While diffusion and innovation exist in their own rights, they also manifest in technology and its transfer; technology and its transfer embody innovation in thought, process, and activity; diffusion is the spread component of the transfer process; technology and TT, innovation, diffusion, and innovation diffusion all share a close relationship; technology could be described as innovation; TT involves innovative practices which generate more innovations and practices through which innovations spread or diffuse into society; technology is more than the artefact; and technology is made up of different bodies of knowledge and their related techniques, contraptions, contrivances, machines, and material and immaterial things, among others. As a subconcept of technology transfer, which is viewed here as a systemic concept, adaptation borrows from the concepts of innovation and diffusion. Furthermore, due to the breadth of applicability of the adaptation concept, it manifests some features and attributes of other concepts and theories that are relevant to TT. In this case, adaptation is framed by technology innovation, technology diffusion, and diffusion of innovations. Moreover, innovation and diffusion exist in other knowledge areas apart from technology. As such they are not only defined within TT alone. However, they frame adaptation within TT, as presented in **Figure 1**.

In the research, analogies were drawn from most of the intellectual traditions of the adaptation concept. The psychology dimension relates to issues of perception of CSM level participants in construction. Such perceptions result from interaction with ICT, with other humans through ICT, and with other humans who use ICT. There is also the issue of perception of CSM process needs and utility in the ICT. The biology and anthropology traditions relate to the need to survive, remain competitive, and be technologically advanced for CSM people and construction industry organisations. It also speaks to the mutual adaptation

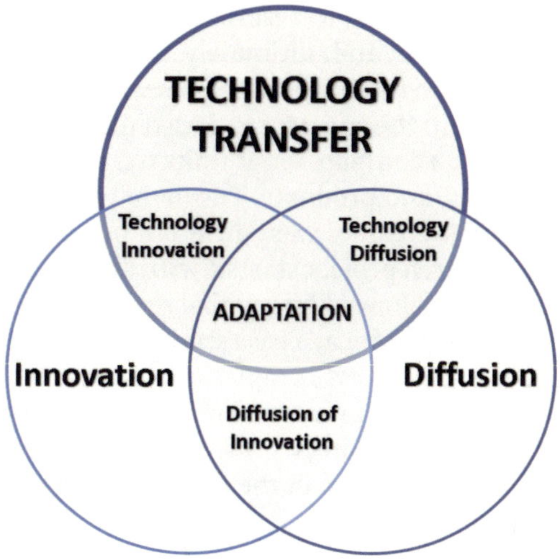

Figure 1.
Framing of the adaptation subconcept of technology transfer [38].

that occurs in the engagement with technology, especially ICT. The sociology and geography traditions relate to the influencing environments, which would impact on the nature of adaptation of ICT in CSM. Following the analogies, the adaptation concept identifies 'structural or functional similarities in otherwise dissimilar situation of things' [36]. This description refers to the identification of opportunities in recent ICT, for weaknesses/inefficiencies/lapses in the dissimilar situation of CSM process. Hence, attributes of the research concept include that in recent ICT, there are utility, structural, and functional characteristics, and features, which are referred to as ICT potential utility. The exploitation of ICT potential utility for CSM through adaptation involves specific steps. Recent ICT is assessed to identify potential utility for possible process enhancement in CSM. The relevant application of such potential utility could be via direct usage, through a degree of modification, or integration of features into a hybrid system. It could be used as a component of other innovative products, and in areas of relevant application. CSM is investigated to compare the available ICT and its transfer. The areas of need in the CSM process, which could be improved by exploitation of ICT potential utility, are determined.

Following the preceding discussion, constructs generated from the study include time frame, which refers to the time frame to complete the adaptation process; unit of adaptation, which is about the key agency in the adaptation; type, which refers to the extent, aspect, and depth of adaptation; pathway, which refers to the natural route of adaptation based on the nature of required/initiated adaptation; and process, which is focused on the source, pattern, end point, and location of adaptation. These were reduced to measurable attributes or variables and examined through a bespoke research methodology.

3. Research design for the chapter

The overall methodological dimensions for the wider research project could be described as a mixed method [39] and multistage research design [40], which utilised quantitative and qualitative data. A fundamentally descriptive purpose was adopted, which started with an exploratory approach and ended with an evaluation.

Essentially, the subject was explored, to describe it in full; then further exploration led to confirmations, evaluations, and, ultimately, a final description. Multiple sources were used, according to the stages in the research design and in compliance with the research concepts and the constructs. Based on the research as determined through preliminary studies, a complex set of units were studied: construction sites, practitioners, manufacturers, and products. The main unit of analysis in each case was ICT in CSM, though the outcomes were different for each stage.

Investigations in the research project started with an exploratory single project, multisite case study. This was followed by a global survey of relevant ICT, a national survey of practitioners in South Africa, a case study of multiple project sites, and a final ICT product analysis. Purposive samples of construction sites, ICT-based products, and construction project management practitioners were chosen. The tools used include observation and interview schedules, content analysis schedules, and questionnaires. MS Excel was used in the initial data analysis. Other analytical tools were employed subsequently.

4. Summary of results and findings and evaluation of hypotheses

In the research project, the first stage was a case study of a highly critical single project with multiple sites, presenting rich baseline information for the research. The highly limited occurrence of ICT, appreciable occurrence of weaknesses in CSM, and apparent lack of awareness of some opportunities in recent in ICT were indicated [38]. The global ICT survey results showed a significant indication of a continuously growing list of ICT products, which are relevant to CSM. There is variety for adequate choice to be made for relevant needs. The products exist locally and internationally. They are accessible through the Internet, physical representation by various manufacturers, and a large population of local vendors. This stage was used in [41]. Product types which are well adapted for CSM, such as mobile hand-held devices and geographic surveillances, have appreciable variety. Such significant adaptation and variety highlight the possible influence of usability and user acceptance, which is a function of the extent of adaptation of the product. Such products would continue to experience further adaptations, towards user demands. It would also follow that the more use an ICT product line experiences, the more transformation the product would experience towards user requirements. In comparison, the national practitioner survey indicated a very low degree of innovativeness in the exploitation of ICT potential utility among practitioners. Results showed that while digital surveying equipment was highly utilised, laser ranger and measurer which are directly related to surveying were not comparatively utilised by site managers on site. High usage of mobile hand-held devices occurred with very low usage of wireless technologies and services. This stage was used as empirical basis for [42]. There was appreciable contrast between knowledge and awareness and utilisation of various ICT types. Apparently, knowledge and awareness of ICT did not necessarily translate into usage within CSM. Similarly, self-assessment of skills for such ICT and some depth of utilisation did not translate to utility of the same ICT in SM by practitioners.

The practitioner survey was also used to explore the experience of barriers to ICT uptake from participants. Network service problem was the highest indication. Results of comparative analysis with findings from the literature review showed adequate correlation in terms of the severity of barriers. Essentially, major challenges seem to emanate from technology and management support and the knowledge adequacy of the practitioners. This stage was used as empirical basis for the study presented in [43]. The fourth stage was a case study of four purposively selected construction sites. Content analyses, walk-through observations, and informal interviews and evaluations were used for collecting data. The occurrence of lapses in CSM and occurrence of relevant ICT, and its utilisation, were the areas of exploration. This stage was used in [44]. Findings indicated appreciable occurrence of management shortcomings or lapses on sites, largely due to human limitations, general management strains and inefficiencies, and the human factor. The observed lapses demonstrate strains on site management, which constitute process need areas [38, 44], that can be addressed by adapting potential utility in available ICT. The issue of knowledge was also highlighted. The last stage was the ICT product study. While determining the existence of ICT with such adaptability was central to the research, the main counterpart was assessing the utility of this adaptability for CSM. Therefore, a purposive sample of ICT products was analysed, through physical observation and content analysis of manufacturer's technical documents and online reviews. This stage was used to further substantiate the existence of potential utility in recent ICT that are relevant to CSM.

4.1 Findings and discussion

Findings from the multistage research project were articulated into publications and additional research and publication of more research papers and articles. The output includes publications based on core empirical data from the research, except for [38], which is the unpublished thesis, and [45–47], which are co-authored with research students. In the latter case, the work was based on the focus, theoretical support, and supervisory guideline, from this research project. See **Table 2**.

Between 2012 and 2013, the first set of core empirical data was used, looking at user awareness and user perception of challenges and effectiveness of ICT [45, 48, 49]. Results highlighted limited knowledge and technology support, but a perceived process improvement and better project experience from adoption of ICT. Extant

Reference	Research output	year
[48]	ICT in site management process in South Africa	2012
[38]	Main research report (unpublished Thesis)	2012–2013
[49]	Investigating barriers to ICT adoption in Site management: a pilot study in South Africa	2013
[45]	Participants' perceptions on investment in ICT and project outcomes in South Africa	2013
[42]	Exploring the knowledge function in the adoption of ICT in site management in South Africa	2014
[46]	Use of ICT-based systems in site security management: a South African study	2014
[41]	Market availability of information and communication technologies and their adoption in site management in South Africa	2017
[47]	ICT in the training of South African bricklaying operatives: a pilot in the Greater Johannesburg area	2017
[43]	Exploring challenges to ICT utilisation in construction site management	2018
[44]	Process need areas and technology adoption in construction site management	2019

Table 2.
Historical perspective of empirical output from the research project.

literature in [50] also indicated the quality of the technology as a strong determinant of ICT use. A causal relationship between ICT use and project performance was also established. In [42], the function of knowledge [29] was explored further and [46] was focused on site security management as an aspect of CSM. From the results, it was determined that knowledge-related factors could be indirectly influencing other challenges, such as lack of management support and ICT requirement analysis and input in contracts. There were indications of possible patterns according to age group and experience. In [42] it was concluded that '...the extent of awareness, skills and working knowledge has not approached the adequacy level to increase adoption of ICT in site management appreciably'. The need for more awareness is highlighted in the recent literature such as [51]. Skills and training, which are knowledge-based, were also highlighted in [52]. However, one significant finding was that knowledge did not translate to adoption and use in every case. In 2017, the global ICT survey data was compared to data on awareness and utilisation of relevant ICT in CSM [41]. In addition, a unique area, training of bricklayers, was used to collect and test current data in terms of market availability of ICT and its occurrence [47]. Results show proof of local availability and access to recent ICT but a relatively poor level of awareness among practitioners and training providers for building trades. The impact of knowledge-related factors was further highlighted. In 2018, the studies on barriers to ICT utilisation were taken forward in [43]. Existing studies, various groupings, and different terms of framing were explored internationally and locally. Results generated the current framework for naming, categorising, and understanding challenges to ICT in CSM, which is supported by literature such as [53, 54]. The possible compounding effects of knowledge, technology, and management support factors were highlighted. In extant literature such as [53], the need for management support and increased availability of technology was highlighted. In [43] emergent patterns of identification of challenges according to age and work were indicated. End user response was identified as highly critical for success, highlighting user acceptance, usability, ergonomics, ease of use, and 'adaptability of the ICT to different related uses' [43], among others. Such issues are supported in literature such as [54, 55]. In 2019, the construct of process need areas [38, 44] derived from the research was also taken forward. Recent literatures and relevant concepts and theories were examined and applied. A pluralistic approach of using five theories/concepts was employed to explain the results, highlighting the need for a more comprehensive framework. It was again noted that acquisition of technology and utility of the technology's adaptability are different issues with varying impacts. While technology was embraced, the exploitation of inherent potential utility was not necessarily occurring in the final use. Results highlighted the need for process improvements in CSM, the limitations and inefficiencies in CSM, and the benefits of ICT for CSM. Findings are supported in extant literature such as [56, 57].

4.2 Summary of the evaluation of the research hypotheses

Following the discussion of findings from the research, five hypotheses were generated, to test the emergent ideas. The hypothesis evaluation process was tailored to the research methodology. Thus, a corresponding multistage approach was used. The process involved breaking down each hypothesis into subhypotheses, and they were evaluated individually, using data from various stages in the research methodology. Results were then outlined in summary as evidence for evaluating the main hypothesis, using the rules of evidence for evaluating the hypotheses. Essentially the null hypothesis in each case is rejected if the summary of evidence significantly supports the claim in the corresponding hypothesis. Otherwise, the null hypothesis is not rejected. To analyse the data sets for evaluation of the hypotheses, the following were used: descriptive statistics, inferential statistics, content analysis, and narratives. The procedures included Z test of proportions [58, 59] and chi-square test of goodness of fit [58]. Initially MS Excel was used as the main analytical tool. Subsequently, other tools such as Stata, ATLAS.ti, NVivo, WordStat, and Qualtrics were added. Frequency and contingency tables, means and proportions, and chi-square tests were used for most of the analysis. In addition, the Z test of proportions was used for the ICT and practitioner surveys.

The first hypothesis, HP1, states that there is limited knowledge of recent ICT-based goods and services that are commercially available, due to lack of awareness and usage in CSM. HP1 was broken down into two subhypotheses, HP1-a (assessing the commercial availability of recent ICT which are relevant to the SMP) and HP1-b (addressing the issue of limited knowledge of recent ICT among practitioners in the SMP). Inferential statistics applied to test availability of recent ICT products indicated that majority of products (above 75%) are available locally and internationally, at 5% level of significance with a minimal 2% risk of error. Secondly the identified ICT types have variety in terms of 85 brands and 439 subbrands, making up a total of 1635 products. Thirdly out of a total possible availability value of 2, all product types had values of 1.8 or above. Similarly, significant occurrence of limited awareness of recent ICT was established from the stage 3 practitioner survey data. Less than 50% of respondents indicated awareness for up to 8 out of the 17 original ICT types considered. Less than 50% awareness values were indicated for the majority of the products (18 out of the 34 types). Furthermore, stage 1 case study indicated that only 3 out of the 35 ICT types under consideration were observed for the 3 sites visited. Moreover, in stage 4 case studies, less than 50% of ICT types under consideration were observed for the 4 sites visited. The evidence supported the commercial availability and range of choice in the majority of recent ICT, for CSM; significantly limited awareness and usage of recent ICT among CSM practitioners; and lack of on-site usage of recent ICT in CSM. Therefore, HP1 is supported.

The second hypothesis (HP2) was stated as follows: The adaptation of recent ICT for the SMP is suboptimal due in part to limited knowledge and management and technology support. HP2 was evaluated through two lenses, namely, HP2-a (suboptimal adaptation) and HP2-b (stated barriers to ICT in the SMP). The summary of evidence showed that the evaluation of HP1 established the limited knowledge of recent ICT for CSM. In terms of adaptation, descriptive analysis of practitioner survey data showed the highest indication of ICT awareness as less than 50% of the highest possible value. Other attributes/variables of ICT adaptation such as use, skills, and on-site use were all under 23% of their highest possible values. Aggregated values indicated a grand mean of 75.35 out of a possible 290. Thus, suboptimal adaptation of recent ICT in CSM was evidenced. Inferential statistics indicated a lack of adaptation of ICT in CSM at a statistical significance of more than 65% or 0.65 of the highest possible sample mean. The case studies evaluated in HP1 and their emergent narratives confirm the low occurrence of ICT in CSM. The narratives also show a strong indication of limited management support. In addition, results from exploring the challenges to ICT in CSM support the hypothesis claim of the relatively higher severity of limitations in technology and management support. Moreover, narratives from stage 4 case studies show strong indications of limited knowledge of potential utility in recent ICT. As such within the research scope, adaptation of recent ICT for CSM was found to be suboptimal, due in part to limited knowledge and management and technology support. Therefore, HP2 is supported.

The third hypothesis (HP3) was stated as follows: Increased adaptation of recent ICT could enable ubiquity and real-time operation of the site information management and communication system. HP3 was evaluated through HP3-a (availability of recent ICT and their relevance to information management and communication in the CSM) and HP3-b (occurrence of lapses in the SMP due to poor information management and communication, which could be addressed through the utility of recent ICT). For the summary of evidence, evaluation of HP1 and HP2 established the availability of recent and relevant ICT products in HP1. The global ICT survey, final product study, and mapping of utility for CSM performed in the research were used to confirm that the majority of the identified ICT are relevant to IM&C. Evaluation of HP2 established the lack of occurrence and suboptimal utilisation of ICT in case study sites and also linked the indicators to suboptimal adaptation of ICT in CSM. Furthermore, the case studies highlighted lapses in site IM&C. This is supported by inferential statistics from the practitioner survey data. The evidence supports the notion of appreciable availability of recent ICT which are relevant to IM&C; the occurrence of lapses in site IM&C; and the occurrence of suboptimal adaptation of relevant ICT to site IM&C. Therefore, HP3 is supported.

The fourth hypothesis (HP4) was stated as follows: Increased adaptation of recent ICT for site H&S management could improve real-time monitoring and reporting. The evaluation of HP4 was broken into three areas, namely, availability of recent ICT and their relevance to site H&S; lapses in site H&S; and suboptimal utilisation of recent ICT in CSM. In terms of evidence, HP1 established the availability of recent and relevant ICT products in HP1. HP3 established the primary relevance of up to 20 ICT items to site H&S and the relevance of recent ICT to site IM&C, which form the basis for HP4. HP2 established the lack of occurrence and utilisation of ICT in case study sites. The same evaluation also linked the two indicators to suboptimal adaptation of ICT in the SMP. Furthermore, results of the case studies provide evidence of lapses in site H&S and lack of occurrence of relevant ICT and their utilisation. HP1, HP2, and HP3 strongly support the claim of suboptimal adaptation of relevant ICT in site H&S. The evidence supports the existence of appreciable availability of recent ICT which are relevant to site H&S management; occurrence of lapses in site H&S; and suboptimal adaptation of relevant ICT to site H&S. Therefore, HP4 is supported.

The fifth hypothesis (HP5) was stated as follows: Increased adaptation of recent ICT for on-site materials logistics could result in efficient and dynamic logistics management. HP5 was also broken into three areas, namely, availability of recent ICT and their relevance to site materials logistics; lapses in site materials logistics; and suboptimal utilisation of recent and relevant ICT for site materials management. In terms of evidence, HP1 and HP3 were used to establish the availability of recent ICT products and their relevance to site IM&C, which forms the basis for HP5. HP3 and further mapping of ICT potential utility established that an appreciable proportion (22/34) of recent ICT is relevant to site materials logistics. HP2 established the lack of occurrence and utilisation of ICT on case study sites and linked the indicators to suboptimal adaptation of ICT in CSM. The case studies also present an evidence of lapses in site materials logistics and lack of occurrence of relevant ICT and their limited utilisation SMM. This is supported by HP1 and HP2, on the claim of suboptimal adaptation of relevant ICT in SMM. As such HP5 is supported.

Following the derivation of findings, synthesis with literature, and evaluation of generated hypotheses, it was possible to build appreciable understanding of various attributes/features of the adaptation concept of ICT transfer to CSM. The understanding is presented in the next section.

5. An understanding of the adaptation concept of technology transfer of ICT in construction site management

Part of the aim of this research was to build an adaptation-based understanding of the transfer of ICT to the CSM. This section presents an exploration of the attributes, constructs, and concepts, the development of their frameworks, and the articulation of a basic flow model and a detailed conceptual model to explain the nature of the adaptation concept of TT, as it applies to ICT in CSM, and construction in general. In addition to process need areas and ICT potential utility, other derived constructs include time frame of adaptation, unit of adaptation, type of adaptation, pathways, and process of adaptation.

5.1 Concepts/constructs generated from the research

The construct of time frame is based on the time factor in diffusion of innovation theory and the innovation decision process. For any outcome and process of adaptation, there is a time frame. The time frame could be immediate in the case where the technology's usefulness is identified simultaneously with awareness of it. Then, the identified potential utility is immediately exploited. There could however be a considerable time lapse of short term or long term. Short term would occur within one construction project event. Long term would occur during multiple project events and probably at an organisational scope or level.

Unit of adaptation as a construct refers to the key agent of the adaptation. Basically, it asks the question: Who is implementing the innovation? The answer could be the individual end user, the work group within a project, the project team/group, or the project organisation. There is also a possible third case where it is initiated by the individual, work group, or project team but adopted and powered by the organisation, all within one process of adaptation. A further possible scenario is the case of adaptation by a project stakeholder other than the contractor, such as client, design team, or major supplier. However, they are viewed as being within the project team or organisation. The unit could also be an external party, while the outcome is meant for construction.

The type of adaptation is comparable to the concept of hard and soft technologies, which has found popularity in the climate change remediation discourse. It refers to the mode taken in adapting technology. Hard adaptation mode is the type that requires modification or fine tuning of technology prior to its usage in the intended environment. The example would refer to physical hardware modification and software re-engineering. Soft adaptation mode refers to a degree of minimal alteration, but it relates more to idea components in innovation diffusion. It would therefore not involve any re-engineering of hardware or software. The innovation is on ideas, not involving physical things. It could also be on an idea on which a product is based. Such adaptations may eventually result in hard adaptation.

The pathway construct is related to the type construct, in that it traces the routes for hard and soft types of adaptation. For pathways, hard adaptation involves hardware adaptation and software adaptation. Soft adaptation could be by ready adaptation, usage adaptation, or context adaptation. Ready adaptation (adoption and use of a well-adapted innovation) is based on the view that adaptation is both a process and an outcome. Therefore, activating the process and acquiring an outcome of the process both fall within the description of adaptation. An example would be the need for remote communication with someone on site and the task-technology fit [60] of the mobile phone which is essentially obvious. Usage adaptation means adapting to a different use from the originally intended use. In one of the case studies in the project, a service provider used a laboratory acidity tester to check the identified groundwater in order to determine if it was portable water leakage or otherwise, using the acidity level. Context adaptation involves adapting from a different intended/primary usage environment to another. The preceding example is also relevant in this case. This category includes adaptation from one sphere, or discipline of origin, to another. The use of laser and infrared scanner in construction and the use of telepresence would be good examples.

Process construct refers to the process of ICT adaptation. It could be completely internal, where it is initiated and completed within a project site/event (internal process). It could also be internally initiated but completed externally in another sector, due to capacity requirements (internal-to-external). In this case the external sphere may be beyond the CSM situation, beyond the immediate project organisation, and at an industry or sectoral level.

To further understand the application of the adaptation concept in this context, the time frame matrix of ICT adaptation is presented in **Table 3**. The time frame construct is used as a categorisation frame, to understand the dynamics of their systemic relationships. While the table is not exhaustive of all possibilities, it gives a clearer sense of what occurs in ICT transfer in construction, from the adaptation perspective, according to time frame.

From **Table 3**, hard adaptation refers to physical or software modification of product, while soft adaptation refers to adaptation without physical adjustment to the product specifications. Hard adaptation could be minor or major. For pathways, there are hardware adaptation and software adaptation, specific to ICT. For process, there is also the case of externally initiated and completed adaptation by an external party, but within the CSM (external process). An example would be where a relatively temporal participant who is external to a project identifies a need and adapts technology to address the need, due to the very limited project exposure. Adaptation could also be initiated by an external party but assumed and completed by site management within CSM (external-to-internal). An example would be where a known need in CSM is addressed through adaptation of technology, by an interested party, who is external to the project, or industry.

Long-term adaptation processes would be the domain of the industry and organisation system levels. This is the realm of strategic alignment and

Adaptation construct	Immediate	Short term	Long term
Unit	Individual	Organisation	Industry level
	Work group	Project level	Organisation
Type	Hard adaptation (minor)	Hard adaptation (major)	Hard adaptation (major)
	Soft adaptation	Soft adaptation	Soft adaptation
Pathway	Ready adaptation	Ready adaptation	Ready adaptation
	Context adaptation	Context adaptation	Hardware adaptation
	Usage adaptation	Usage adaptation	Software adaptation
		Software adaptation	
Process	Internal process	Internal process	Internal process
	External process	Internal-to-external	Internal-to-external
		External-to-internal	External-to-internal

Table 3.
Time frame matrix of ICT adaptation in CSM [38].

organisational management decisions and implementation. The organisational level transcends the short term and long term and enables effective translation of strategy into implementation. Between short term and immediate occurrence, the division could have grey areas because there are more individuals influencing the dynamics of adaptation in these divisions. However, hard adaptation would not occur in the immediate as it requires some degree of preconditioning of the ICT before use. It is argued that soft adaptation could occur within the short term and long term, depending on how sensitive management is to their needs and relevant opportunities. Similarly, usage adaptation could occur in the short term, but context adaptation pathway and internal-to-external adaptation process would typically fall into short term and long term. Ready adaptation could also be long term, in the case of management taking considerable time to make decisions. Furthermore, protracted implementation processes would inevitably result in long-term adaptation. Moreover, it is possible that some forms of adaptation could outlive a project and continue in other projects, at an organisational level. A basic conceptual model of the dynamics of adaptation of ICT in CSM is shown in **Figure 2**, using the cluster of constructs and attributes in their subenvironments and the flow of the adaption process within this context.

Time is used as one major environment of the adaptation process, to emphasise the feature of duration. However, there are other environmental factors such as compatibility requirements [61], which influence the adaptation of technology generally. Arguably, compatibility requirements would include monetary compatibility, which refers to the financial demands of transferring the technology; materials compatibility, which refers to the receiver being able to provide compatible material resources for the functioning and maintenance of the technology; production level compatibility, which refers to the technology being able to function productively and/or being replicated in the receiving environment; infrastructure compatibility, which refers to the availability and reliability of the enabling infrastructural environment; social and political compatibility, which refers to factors such as culture, national, and political interests and skills; and ecological compatibility, which refers to the environmental impact of the technology. It also refers to the technology being transferable in an environment [61].

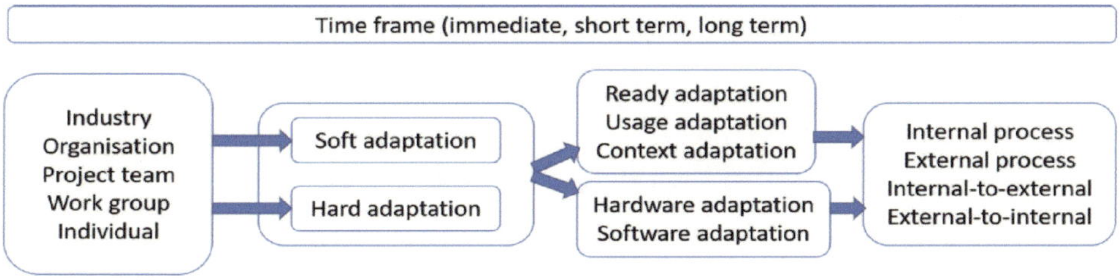

Figure 2.
Basic conceptual model of the dynamics of adaptation of ICT in CSM [38].

From the findings and discussions thus far, and developments in **Table 3** and **Figure 2**, a detailed conceptual model of the nature of ICT adaptation in CSM, and in the construction projects, is proposed in **Figure 3**.

The attributes are organised in their subsystemic relationships, and their systemic relationships, using the unit, time, process, type, and pathway constructs as bases. The unit construct is primary, followed by time. The type of adaptation and its process and pathway then emerge. The project-related situation, which is the unit, could be an individual, work group, project team, firm or organisation, or the industry.

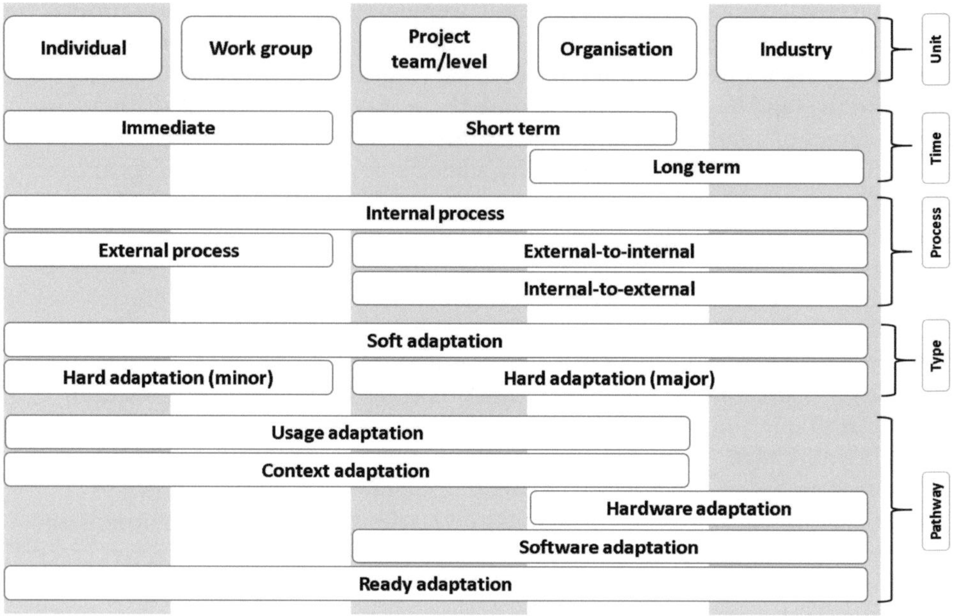

Figure 3.
Model of adaptation dynamics of technology transfer in CSM [38].

6. Conclusions

This chapter presented a research on the uptake of ICT to enhance construction site management process, using the South African context, through the lens of the adaptation subconcept of TT. Essentially, the exploitation of the adaptability of recent ICT to improve CSM was investigated. Through the series of studies leading to the chapter, the common thread has been defined by a collection of concepts/constructs, which embody the essence of the research: information and communication technology; technology; technology transfer; diffusion; the site management process; and adaptation. A deeper understanding of innovation [62], technology acceptance and adoption [63], moderating effects on adoption [64], and the nexus between user-centeredness and technology acceptance [65], was derived. The central discussion is the exploitation of adaptability in recent ICT, as opposed to the adaptability itself, which is established in extant literature. This approach addresses a management problem rather than an application problem.

6.1 Overview of the research

The research is predicated on the problem of suboptimal adaptation of potential utility in recent ICT to improve the CSM. Three aspects of CSM were focused, namely, H&S, SMM, and IM&C. Though various research methods have inherent weaknesses, a robust methodology with adequate rigour was developed to address such weaknesses. The research was conducted through a mixed method, and multistage research design, which generated results and hypothetical statements. The hypotheses alleged suboptimality in the exploitations of ICT potential utility to improve CSM in the focus areas, limited knowledge of available relevant ICT, and the existence of some key challenges to uptake of ICT in CSM. The literature review on technology, ICT, and the construction industry pointed to the sheer capacity of modern technology, especially ICT. It also highlighted the applicability, scalability, and pervasive nature of ICT. These attributes substantiate the adaptability of recent ICT, which could be usefully exploited to achieve far-reaching advantages in construction, especially CSM. Following the deductions from literature review and conceptual basis for the methodology, the analysis of data showed a strong evidence of the existence of gaps identified in the literature. It also presented more telling indications in the local context. Suboptimal exploitation of the adaptability of recent ICT for CSM was highlighted. Issues of limited awareness, limited working knowledge, and limited management and technology support were highlighted. The gravity of these specific issues and their synergistic impact on ICT adoption in CSM was strongly supported by the evaluation of the hypotheses. The overarching deduction with regard to the research problem is that the hypotheses as stated are supported. Within the limits of the research project, the results strongly substantiate the claim that the exploitation of the adaptability in recent ICT to improve the CSM is suboptimal. This flies in the face of the proliferation and availability of such technologies and the performance of other industries in that regard.

6.2 General conclusion to the research

Findings from the literature review are that ICT adoption in CSM is inadequate and inexpedient, falling short of potential benefits for construction. The literature review for this research revealed the need of humans for technology to enhance their natural processes and take full advantage of their natural environment. This need has evolved and extended to every aspect of human life including construction and CSM. Literature also revealed that construction has developed largely through technological innovations. Particularly the industry in recent history has grown by exploiting the adaptability of technology originating from other sectors such as ICT. Through innovation, rather than reinvention, many technologies have been successfully adapted to construction in areas such as CSM. Literature revealed the capacity and pervasiveness of ICT in modern life and its relevance to the enhancement of CSM. Nevertheless, adaptation of ICT to construction and CSM depends appreciably on factors including awareness and working knowledge and technology and management support. CSM contends with challenges of limited human capacity, overstretching of management, and blind spots in overseeing the site constitutions and activities. This leads to lapses in various areas of CSM such as IM&C, H&S, and SMM. Thus, the SMP has many potential beneficiary areas for adaptation of recent ICT. However, literature from the study has also shown that the construction industry in general falls behind other industries in adapting potential utility of recent ICT to improve CSM. While not entirely resulting from the lack of adequate ICT, lapses observed in CSM case studies demonstrate weaknesses in site management, which cannot be effectively addressed through increase in personnel as observed. In practice, not all designated personnel are on site at the same time, due to unsustainability of such measures. Furthermore, there is currently no known substitute to ICT in addressing requirements of ubiquity and real-time capacity.

Field research results pointed to underutilisation of innovations located in recent technology. The main causative factors include inadequate knowledge of potential utility of recent ICT and a lack of effective approaches to technology integration and management support. Results highlight the fact that all possible units of adaptation agency can initiate and drive the required innovation. Beyond the primary usage of products, there is more utility offered in their designs and specifications. It is therefore important that usage of ICT goes beyond what is immediately discernible as practical usage. Exploiting potential utility in recent ICT requires innovative thinking and approach in their use. The need identified here applies to the individual, group, project team, organisation, and industry. Thus, while the demand for construction products and better delivery persists, putting construction under more pressure, the ICT sector continues to produce relevant products, services, and ideas which can be usefully exploited by construction. Useful exploitation of such output would enhance construction performance in many areas.

Through a customised approach, the research established foundational understanding of the adaptation of ICT in CSM. Contributions from the research include the time frame matrix of adaptation of ICT in CSM, the basic flow model of adaptation of ICT in CSM, and the detailed conceptual model of adaptation of ICT in CSM. Thus, the research contributes to the following bodies of knowledge: technology transfer, innovation, ICT in construction, user experience of ICT in construction, adaptation of ICT in construction, and construction project management. The proposed model is arguably relevant to the exploitation of the adaptability of ICT in other project-based systems and environments, beyond CSM, construction project management, and the construction industry.

The work presented in this chapter is the conclusion to a research project which was first proposed in 2007 and essentially started in 2008, generated a doctoral report in 2013 and additional publications up to December 2019. Through 19 publications, the attributes and dynamics of the adaptation concept for TT of ICT in the CSM have been derived, framed, and modelled. The outcomes at this stage are still not exhaustive of all possible scenarios. Future studies should therefore focus on the uptake of ICT/IT in other aspects of construction, user experience (UX) issues, other levels of construction project management, and comparison of the adaptation concept with other relevant concepts, and the use of contributions from the current study, to work on ICT based solutions for various needs in construction. Such studies should also aim to develop a deliberate/formalised approach to the transfer of ICT in construction, using the natural patterns discovered in the research, emphasizing user centeredness and user experience.

Author details

Aghaegbuna Obinna U. Ozumba[1*] and Winston Shakantu[2]

1 University of the Witwatersrand, Johannesburg, South Africa

2 Nelson Mandela University, Port Elizabeth, South Africa

*Address all correspondence to: obinna.ozumba@wits.ac.za

References

[1] Lahiri R, Ding J, Chinzara Z. Technology adoption, adaptation and growth. Economic Modelling. 2018;**70**:469-483. DOI: 10.1016/j.econmod.2017.08.026

[2] Bauer ST, Flagg JL. Technology transfer and technology transfer intermediaries. Assistive Technology Outcomes and Benefits Journal (ATOB). 2010;**6**:1. Available from: https://www.atia.org/wp-content/uploads/2015/10/ATOBV6N1.pdf

[3] Lane JP. Understanding technology transfer. Assistive Technology. 1999;**11**(1):5-19. DOI: 10.1080/10400435.1999.10131981

[4] Ojiako U, Froise T, Shakantu W, Ozumba AO, Alasdir M, Chipulu M. Building information modeling (BIM) as a collaborative tool in construction project delivery. In: Proceedings of the UK Academy for Information Systems (UKAIS) Information Systems Comes of Age: 21 Years, Looking Back to Look Forward. Oxford: UKAIS; 2016

[5] Venkatachalam S. Analysis of the challenges and implementation strategies on the adoption of building information modelling in south African architectural firms. Journal of Construction Engineering. 2015;**8**(2):7-12

[6] Ozumba AOU, Shakantu WM. Improving site management process through ICT. In: Proceedings of the CIDB 5th Postgraduate Conference on Construction Industry Development; 16-18 March 2008; Bloemfontein. Pretoria: CIDB; 2008. pp. 22-31

[7] Ozumba AOU, Shakantu WM. Achieving ubiquity in the site management process: A theoretical study of the potential for innovative ICT solutions. In: Proceedings of the RICS Construction and Building Research Conference; 4-5 September 2008; Dublin. London: RICS; 2008. pp. 22-31

[8] Ozumba AOU, Shakantu WM. Improving materials management through utilisation of information and communication technology. In: Proceedings of the SACQSP Research Conference on Construction beyond 2010; September 2008; Johannesburg. Johannesburg: SACQSP; 2008. p. 23

[9] Ozumba A, Shakantu W. Balancing site information and communication technology systems with available ICT skills. In: Proceedings of the RICS Construction and Building Research Conference; 10-11 September 2009; Dublin. London: RICS; 2009. pp. 128-137

[10] Ozumba A, Shakantu W. Information and communication technology – Based application of 'just- in- time' (JIT) to internal logistics on site improving site management process through ICT. In: Proceedings of the RICS Construction and Building Research Conference; 10-11 September 2009; Dublin. London: RICS; 2009. pp. 447-461

[11] Ozumba AOU, Shakantu WMW. Enhancing on-site communications by adaptation of multimedia systems: Looking into the future of construction. In: Proceedings of the SACQSP Quantity Surveying Conference on the Future – What Next?; September 2009; Durban. Johannesburg: SACQSP; 2009. p. 06

[12] Ozumba AOU, Shakantu WMW, Smallwood JJ. Improving people-centeredness in H&S risk management through ICT. In: Proceedings of the CIB TG59 Conference on People in Construction; March 2009; Port Elizabeth. Port Elizabeth: Nelson Mandela metropolitan University; 2009. p. 34

[13] Ozumba AOU, Ata ON, Oburo NC, David N. Cutting edge technology for construction ICT in a developing

country. In: Proceedings of the RICS Construction and Building Research Conference; 2-3 September 2009; Paris. London: RICS; 2010

[14] Ozumba AOU, Dlamin TC, Shakantu WMW. Information and communication technology education within south African built environment schools. In: Proceedings of the RICS Construction and Building Research Conference; 2-3 September 2009; Paris. London: RICS; 2010

[15] Ozumba AOU, Udeh CA, Shakantu WMW. Indigenous iron technology evolution: Lessons from Uzu culture of Awka in Nigeria. In: Proceedings of the CIDB 5th Postgraduate Conference on Construction Industry Development; 16-18 March 2008; Bloemfontein. Pretoria: CIDB; 2008. pp. 22-31

[16] Leung D, Lo A, Fong LHN, Law R. Applying the technology-organization-environment framework to explore ICT initial and continued adoption: An exploratory study of an independent hotel in Hong Kong. Tourism Recreation Research. 2015;**40**(3):391-406. DOI: 10.1080/02508281.2015.1090152

[17] Chen JS, Tsou HT. Information technology adoption for service innovation practices and competitive advantage: The case of financial firms. Information Research. 2007;**12**:3. Available from: http://informationr.net/ir/12-3/paper314.html

[18] Rizk Y, Bhandwalder A, Boag S, Chakraborti T, Isahagian V, Khazaeni Y, et al. A Unified Conversational Assistant Framework for Business Process Automation [cs.AI]. 2020. arXiv:2001.03543v1. Available from: https://arxiv.org/pdf/2001.03543.pdf

[19] Stohr EA, Zhao JL. A Technology Adaptation Model for Business Process Automation. In: Proceedings of The Thirtieth Annual Hawwaii International Conference on System Sciences. Hawaii: IEEE Computer Society; 1997. pp. 405-414

[20] Zhang B, Pavlou PA, Ramayya K. On direct versus indirect peer influence in large social networks. Information Systems Research. Fox School of Business Research Paper No. 17-016; 2017. Available from: https://ssrn.com/abstract=2989057

[21] Milli L, Rossetti G, Pedreschi D, Giannotti F. Information diffusion in complex networks: The active/passive conundrum. In: Cherifi C, Cherifi H, Karsai M, Musolesi M, editors. Complex Networks & Their Applications VI. COMPLEX NETWORKS 29 November – 1 December 2017. Studies in Computational Intelligence. Vol. 689. Lyon, France: Springer; 2018. pp. 305-313

[22] Pensupap V, Walker DHT. Innovation diffusion at the implementation stage of a construction project: A case study of information communication technology. Construction Management and Economics. 2006;**24**:321-332. DOI: 10.1080/01446190500435317

[23] Hardey M. Gender and technology culture: Points of contact in tech cities. Sociological Research Online. 2019;**25**(1):101-118. DOI: 10.1177/1360780419851137

[24] Wyrwicka MK. Technology culture and its interpretation: Research results in Poland. Human Factors. 2010;**21**:178-187. DOI: 10.1002/hfm.20254

[25] Organisation for Economic Co-operation and Development (OECD). North/South technology transfer: The adjustments ahead. Report, OECD Committee for Scientific and Technological Policy, Paris. Paris: OECD; 1981

[26] Drucker P.F. The discipline of innovation. Harvard business Review. November–December 3-8, 1998.

Available from: http://wwwcbpa.louisville.edu/bruceint/drucker.pdf

[27] Lou TF, Li EY. Integrating innovation diffusion theory and the technology acceptance model: The adoption of blockchain technology from business managers' perspective. In: Proceedings of the 17th International Conference on Electronic Business, December 4-8. Dubai; UAE: ICEB; 2017. pp. 293-296

[28] Rogers EM, Shoemaker FF. Communication of Innovations. New York: The Free Press; 1971

[29] Rogers EM. Diffusion of Innovations. 5th ed. New York: The Free Press; 2003

[30] Rogers EM, Singhal A, Quinlan MM. Diffusion of innovations. In: Stacks D, Salwen M, editors. An Integrated Approach to Communication Theory and Research. New York: Routledge; 2006, 2006. Available from: http://utminers.utep.edu/asinghal/reports/emr-singhal-quinlan-june19-07-doi-word-file-stack-salwen%5B1%5D.pdf

[31] van Oorschot JAWH, Hofman E, Halmana JIM. A bibliometric review of the innovation adoption literature. Technological Forecasting and Social Change. 2018;**134**:1-21. DOI: 10.1016/j.techfore.2018.04.032

[32] Lee N, Li S, Shin B, Kwon O. Social comparison, goal contagion, and adoption of innovative information technology. The Journal of Computer Information Systems. 2016;**56**(2):127-136. DOI: 10.1080/08874417.2016.1117374

[33] Simonet G. The concept of adaptation: Interdisciplinary scope and involvement in climate change. SAPIENS. 2010;**3**:1. Available from: https://journals.openedition.org/sapiens/997

[34] Amundson R. Historical development of the concept of adaptation. In: Rose MR, Lauder GV, editors. Adaptation. New York: Academic Press; 1996. pp. 11-53. Available from: https://www.researchgate.net/publication/233820303_Historical_development_of_the_concept_of_adaptation

[35] Edwards P. Adaptation: Two theories. Text and Performance Quarterly. 2007;**Q.27**:369-377. DOI: 10.1080/10462930701587592

[36] Rosenbloom RS. The transfer of military technology to civilian use. In: Kranzberg M, Pursell CW, editors. Technology in Western Civilization Volume II. London: Oxford University Press; 1967. pp. 601-612

[37] Holliday GD. Western technology transfer to the Soviet Union, 1928-1937 and 1966-1975: With a case study in the transfer of automotive technology [thesis]. Washington D. C.: The George Washington University; 1978

[38] Ozumba AOU. Exploiting the adaptability of recent ICT to improve the site management process [thesis]. Port Elizabeth: Nelson Mandela Metropolitan University; 2013

[39] McKim CA. 2015. The value of mixed methods research: A mixed methods study. Journal of Mixed Methods Research. 2015;**11**(2):202-222. DOI: 10.1177/1558689815607096

[40] Gesink D, Filsinger B, Mihica A, Norwood TA, Sarai C, Perez RD, et al. Cancer screening barriers and facilitators for under and never screened populations: A mixed methods study. Cancer Epidemiology. 2016;**45**:126-134. Available from: https://www.sciencedirect.com/science/article/abs/pii/S1877782116302041

[41] Ozumba AOU, Shakantu WMW. Market availability of information and communication technologies and their

adoption in site management in South Africa. In: Proceedings of the ASOCSA Research Conference; 6-8 August 2017; Durban. Durban: ASOCSA; 2017. pp. 300-311

[42] Ozumba A, Shkantu W. Exploring the knowledge function in the adoption of ICT in site management in South Africa. The Journal of Construction Project Management and Innovation. 2014;**4**(2):990-1011

[43] Ozugmba AOU, Shakantu W. Exploring challenges to ICT utilisation in construction site management. Construction Innovation. 2018;**18**(3):321-349. DOI: 10.1108/CI-03-2017-0027

[44] Ozumba AO, Ojiako U, Shakantu W, Marshall A, Chipulu M. Process need areas and technology adoption in construction site management. The Journal of Construction in Developing Countries. 2019;**24**(2):123-155. DOI: 10.21315/jcdc2019.24.2.6

[45] Pedro L, Ozumba A. Participants' perceptions on investment in ICT and project outcomes in South Africa. In: Proceedings of the RICS Construction and Building Research Conference; 10-12 September 2013; New Delhi. London: RICS; 2013

[46] Ozumba A, Mapatha D, Shkantu W. Use of ICT-based systems in site security management in South Africa. Journal of Construction Project Management and Innovation. 2014;**4**(2):912-929. Available from: https://hdl.handle.net/10520/EJC165619

[47] Ozumba AOU, Botjie T, Daya B, Leonard JS. ICT in the training of south African bricklaying operatives: A pilot in the greater Johannesburg area. In: Proceedings of the ASOCSA Research Conference; 6-8 August 2017; Durban. Durban: ASOCSA; 2017. pp. 49-59

[48] Ozumba A, Shakantu W. ICT in site management process in South Africa. In: In Proceedings of the RICS Construction and Building Research Conference; 10-13 September 2012; Las Vegas. London: RICS; 2012. pp. 1189-1197

[49] Ozumba A, Shakantu W. Investigating barriers to ICT adoption in site management: A pilot study in South Africa. In: Proceedings of the RICS Construction and Building Research Conference; 10-12 September 2013; New Delhi. London: RICS; 2013

[50] Ikediashi DI, Ogwueleka AC. Assessing the use of ICT systems and their impact on construction project performance in the Nigerian construction industry. Journal of Engineering, Design and Technology. 2016;**14**(2):252-276. DOI: 10.1108/JEDT-08-2014-0047

[51] Osunsanmi TO, Aigbavboa C, Oke A. Construction 4.0: The future of the construction industry in South Africa. International Science Index, Civil and Environmental Engineering. 2018;**12**:3. Available from: http://scholar.waset.org/1307-6892/10008621

[52] Lew YL, Toh TC, Lim KL, Yan FYY, Yow LP. A study on the constraints of implementing information and communication technology (ICT) in Malaysian construction industry. IOP Conference Series: Earth and Environmental Science. 2019;**385**:1. DOI: https://iopscience.iop.org/article/10.1088/1755-1315/385/1/012005/meta

[53] Ern PAS, Kasim N, Masrom MAN, Chen GK. Overcoming ICT barriers in IBS management process in Malaysia construction industry. In: Proceedings of the ISCEE 2016; MATEC Web of Conferences. Vol. 103. 2017, 2017. p. 03007

[54] Bosch-Sijtsema P, Isaksson A, Lennartsson M, Linderoth HCJ. Barriers and facilitators for BIM use among

Swedish medium-sized contractors – "We wait until someone tells us to use it". Visualization in Engineering. 2017, 2017;**5**:1-3. DOI: 10.1186/s40327-017-0040-7

[55] Eadie R, Odeyinka H, Browne M, McKeown C, Yohanis M. Building information modelling adoption: An analysis of the barriers to implementation. Journal of Engineering and Architecture. 2014;**2**(1):77-101

[56] Fang W, Ding L, Luo H, Love PED. Falls from heights: A computer vision-based approach for safety harness detection. Automation in Construction. 2018;**91**:53-61. DOI: 10.1016/j.autcon.2018.02.018

[57] Dong C, Li H, Luo X, Ding L, Seibert J, Luo H. Proactive struck-by risk detection with movement patterns and randomness. Automation in Construction. 2018;**91**:246-255. DOI: 10.1016/j.autcon.2018.03.021

[58] Remenyi D, Onofrei G, English J. An Introduction to Statistics Using Microsoft Excel. 2nd ed. UK: Academic Publishing Ltd.; 2010

[59] Agresti A. An Introduction to Categorical Data Analysis. 2nd ed. New Jersey: John Wiley and Sons Inc.; 2007

[60] Golizadeh H, Hosseini MR, Edwards DJ, Abrishami S, Taghavi N, Banihashemi S. Barriers to adoption of RPAs on construction projects: A task–technology fit perspective. Construction Innovation. 2019;**19**(2):149-169. DOI: 10.1108/CI-09-2018-0074

[61] Radhakrishna S, editor. International Council of Scientific Unions (ICSU) and committee on science and Technology in Developing Countries (COSTED). Science, technology and global problems, views of the developing world. In: Proceedings of the Kuala Lumpur Symposium; 27-30 April 1979. UK: Pergamon Press; 1980

[62] Taylor SP. What is innovation? A study of the definitions, academic models and applicability of innovation to an example of social housing in England. Open Journal of Social Sciences. 2017;**5**:128-146. DOI: 10.4236/jss.2017.511010

[63] Taherdoost H. A review of technology acceptance and adoption models and theories. Procedia Manufacturing. 2018;**2018**(22):960-967. DOI: 10.1016/j.promfg.2018.03.137

[64] Ameen N, Willis R. An analysis of the moderating effect of age on smartphone adoption and use in the united Arab emirates. Social transformation. In: Proceedings of the 23rd UK Academy for Information Systems (UKAIS) International Conference. Oxford: UKAIS; 2018

[65] Shore L, Power V, de Eyto A, O'Sullivan LW. Technology acceptance and user-centred design of assistive exoskeletons for older adults: A commentary. Robotics. 2018;7:1-3. DOI: 10.3390/robotics7010003

Chapter 7

Recent Developments of Combined Heat Pump and Organic Rankine Cycle Energy Systems for Buildings

Wahiba Yaïci, Evgueniy Entchev,
Pouyan Talebizadeh Sardari and Michela Longo

Abstract

To develop efficient and lower emission heating and cooling systems, this book chapter focuses on interests for the innovative combination of a heat pump (HP) and organic Rankine cycle (ORC) for building applications. In this state-of-the-art survey, the potentials and advantages of combined HP-ORC systems have been investigated and discussed. Past works have examined various combinations, comprising indirectly-combined as series and parallel, directly-combined units, as well as reversible combination configurations. Following describing such arrangements, their performance is discussed. Considerations for optimising the overall architecture of these combined energy systems are pinpointed using these same sources, taking into account heat source and sink selection, expander/compressor units, selection of working fluids, control strategies, operating temperatures, thermal energy storage and managing more variable seasonal temperatures. Furthermore, experimental works present further functional problems and matters needing additional research, and assist to emphasise experimental techniques that can be utilised in this field of research. Finally, from the studies surveyed, some areas for future research were recommended.

Keywords: heat pump (HP), organic Rankine cycle (ORC), energy systems, microgeneration, design, performance, buildings

1. Introduction

Overall, the building sector represents approximately 40% of the final utilisation of energy and 36% of greenhouse gas emissions (GHGs). For meeting international emission targets, there is a requirement for more efficient heating and cooling systems in order to decrease electricity needs while improving system efficiency as well as reliability; this is because such systems involve over 80% of residential heating usage in many nations and particularly in countries with colder climates, such as Canada [1]. In buildings, advanced heating and cooling as well as micro cogeneration technologies can possibly decrease electricity and fossil fuel-derived use via increased usage of renewable energy sources, thermal storage, micro-cogeneration as well as systems with better efficient energy systems. A thermodynamic-based, the organic Rankine cycle (ORC) is a remarkable system that is appropriate for recovering low-temperature heat from different heat sources, such as solar, geothermal or low-grade thermal power sources for cogenerating heat and power. Of noteworthy relevance is its combination with a heat pump (HP), with which it would be able to more efficiently supply heating and cooling, hence decreasing electricity utilisation and generation of pollutant emissions.

The thermodynamic organic Rankine cycle is characterised as a heat and power generation system, which is labelled as organic due to it utilising working fluids and this allows it to operate at lower temperatures, contrastingly to a steam Rankine cycle. This is positive as it allows the system temperatures to operate efficiently at lower temperatures, in this manner permitting small-scale applications.

Figure 1 illustrates a typical ORC cycle. Advanced ORC systems can include additional units such as a regenerator and recuperator units or extra heat exchanger/turbine, alike to the assortment of potential components for a steam Rankine cycle. In the evaporator, the heat source transfers heat to the working fluid, which evaporates and is hereafter pressurised. This working fluid is sent to the expander turbine unit, which converts the high-pressure gas to low pressure gas, converting the work produced from this process into electricity. The low-pressure gas moves in the condenser, where surplus heat that has not been converted into electricity is dissipated to the heat sink, raising the heat sink fluid's temperature and condensing the working fluid. Ultimately, a pump is employed for circulating the working fluid flow in the cycle [2, 3]. Detailed principle and research works on ORC systems can be found in [2–13].

A heat pump has an analogous working operation to that of an organic Rankine cycle, apart from it just supplying heat rather of generating of power. As its name indicates, it finally receives heat from the heat source and transfers it to the heat sink, reducing the heat source temperature and rising the heat sink temperature. A compressor and a throttling valve are employed for aiding drive and enhance the performance of this process, which can, on typical manner, operate in either heating or cooling mode, and this is where both the heat sink and source are interchanged. The efficiency of the heat pump is denoted by its coefficient of performance (COP), defined as the ratio of total heat delivered by the heat pump to the amount of electricity needed to drive the heat pump. An example of the unit's schematic is illustrated in **Figure 2**. One main benefit of this cycle is that it is able of removing heat from a heat source versus a temperature gradient with a

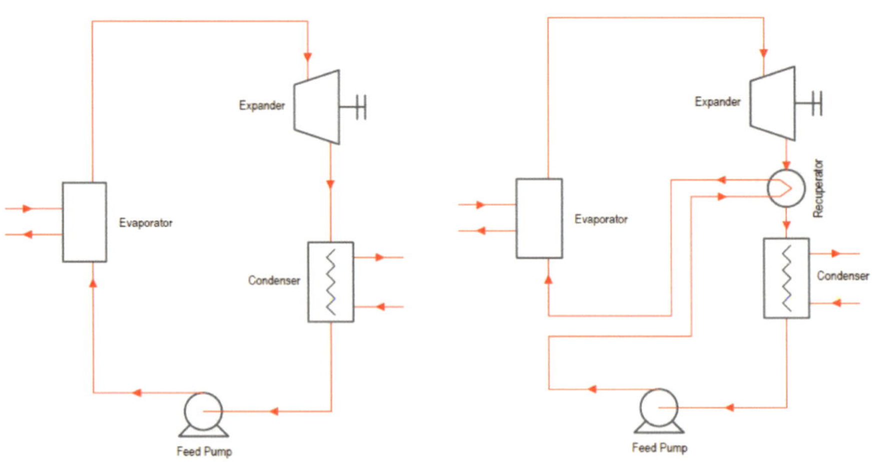

Figure 1.
Schematic of the ORC without (left) and with (right) recuperator.

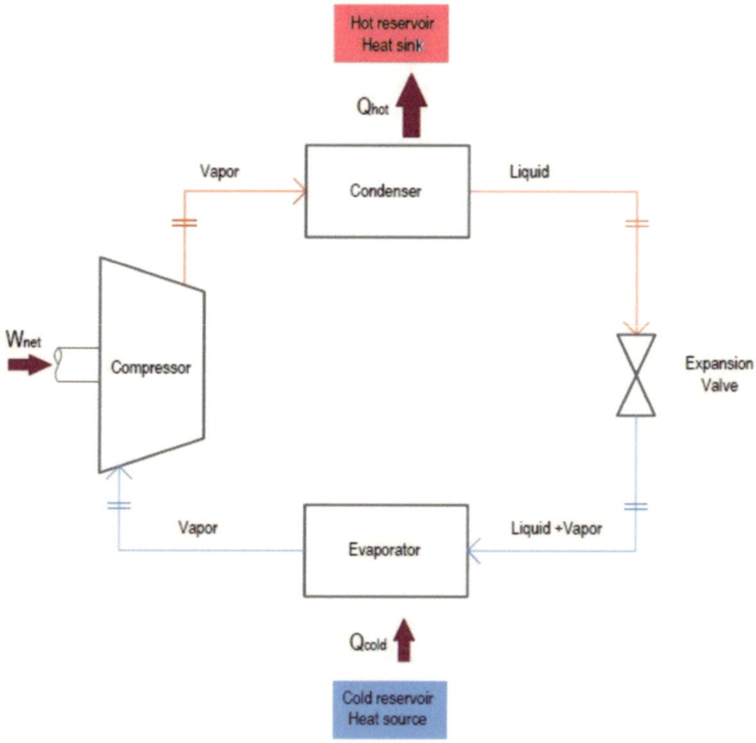

Figure 2.
Schematic of the heat pump.

higher temperature heat sink, enabling it to get to low-grade thermal sources. A heat pump has an assortment of arrangements and configurations available, which are generally out of the extent of this review. Representative heat pumps viable, incorporate a water-source heat pump (WSHP) that include a liquid–liquid heat exchange; a ground-source heat pump (GSHP), and this is fundamentally the same as a WSHP, however the heat source liquid works as a heat transfer medium for geothermal heat exchangers as well; and an air-source heat pump (ASHP) that includes an air-liquid heat exchange, with performance reliant on ambient conditions. Moreover, there are likewise absorption as well as adsorption heat pumps that have increasingly complex operation. For the motivations behind this paper, the heat pump arrangements are optional as against to the overall system arrangement, and thus, will not be examined thoroughly. More detailed works regarding HP systems can be found in [14–29].

In spite of there being various surveys addressing HPs and an ORC system's design, performance and technical-economic studies, there are a lack of reviews focusing on the combined HP and ORC systems for low-temperature and micro co/trigeneration applications. Hence, building upon and extending the work of the authors [30], the aim of the present chapter is to focus on innovative combination of a HP and ORC for use in buildings. In this overview survey, the potentials and advantages of combined HP-ORC systems will be investigated and discussed.

With these objectives in mind, the remaining part of this chapter is organised as follows: Section 2 presents the various types of combined HP-ORC systems. Section 3 elaborates the performance results from the configurations in the Section 2. Section 4 presents further thorough on specific design considerations for optimisation purposes found in important works reviewed. Section 5 summarises findings from experimental works. This later section is followed by a brief overview of some studies on the economic analysis. Finally, the main conclusions drawn in this chapter are provided in the last section.

2. Types of combined HP-ORC systems

This section will give a general overview of various types of HP-ORC systems and discuss related works that are resulted with them.

2.1 Indirectly-combined HP-ORC system

These systems are not directly combined, which will be considered later on in this section. Series HP-ORC systems have comparable connection points to those that are directly coupled, however they have an intermediary loop or device amidst them. An illustration here is a gas-engine HP-ORC system that recuperates engine exhaust gas as ORC heat source as shown in **Figure 3** [31].

Parallel systems provide additional versatility in their mode of operation and are appropriate to be better for regions that have varying temperatures during the year. A study by Li et al. [32] summarised a parallel system to deliver heat, heat storage, as well as power for continental climates, with this system displayed in **Figure 4**. This system included an ORC and a HP, both being able to generate sufficient heat for one household during the cold, or heating, season. A ground heat exchanger (GHE) was utilised as thermal storage, being the hot source that the HP extracted heat from during the heating season and the ORC recharged it during the non-heating season annually.

2.2 Directly-combined HP-ORC system

As the name implies, a directly-combined HP-ORC system is one where the same process unit is shared by the separate HP and ORC units. From the literature, which has been analysed, this takes the arrangement of either a common heat exchanger

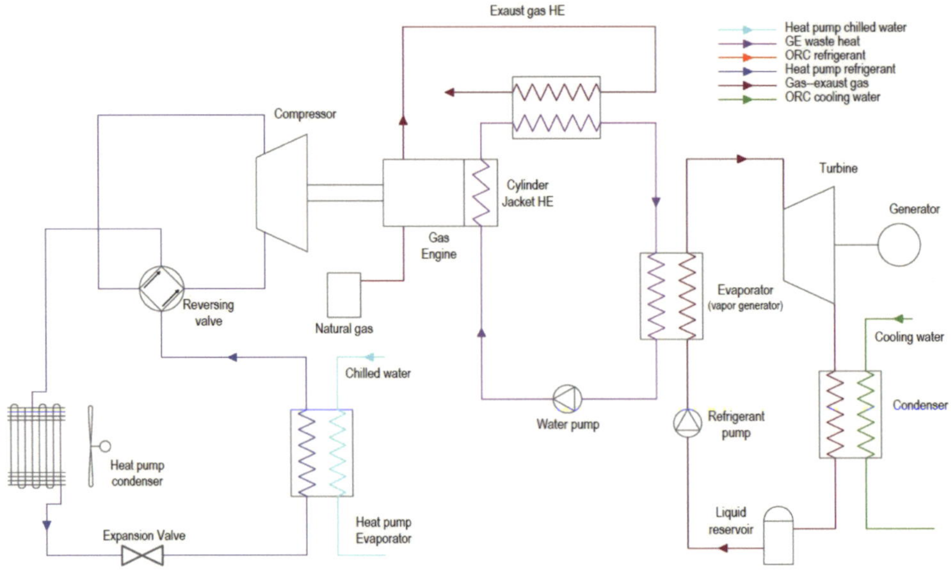

Figure 3.
Schematic of the gas engine HP with the ORC unit for exhaust waste heat recovery.

where the process streams transfer energy to each other with one being the heat source and the other functioning as the sink or serving of a coupled expander/compressor set. An example of the joint heat exchanger HP-ORC system, which was examined by Yu et al. [33] is illustrated in **Figure 5**. The HP unit shares its condenser that is also the evaporator of the ORC. This design is particular where the recovery of waste heat from another process takes place from the ORC evaporator to the HP evaporator and in return to the preheater.

The design from Roumpedakis' work [34] takes a very analogous technique, although this system is constructed in the opposed manner, with the sorption HP obtaining waste heat from the ORC and the heat source for the ORC being a solar thermal loop that underwent assessment for Amsterdam and Athens.

A comparable system to the previously mentioned one can be found in a work by Bellos and Tzivanidis [35], where the absorption HP with working fluid LiBr-H_2O, harnesses the rejected heat from the ORC that is solar parabolic trough thermal based.

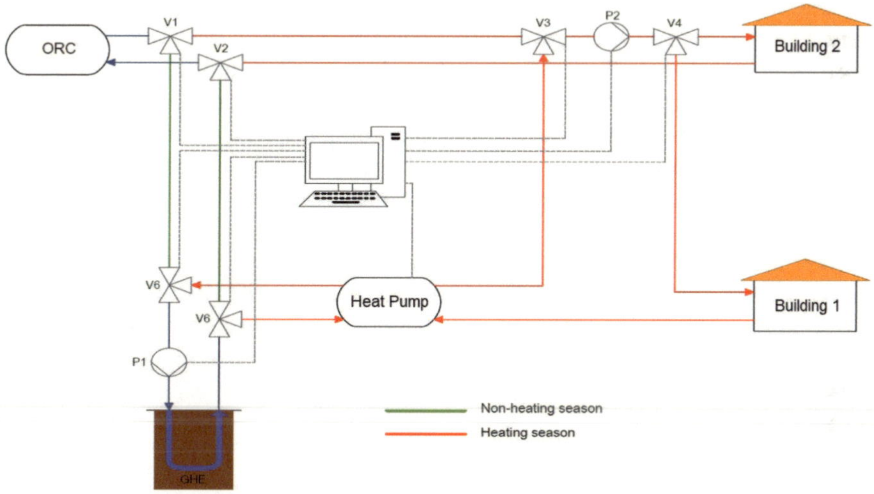

Figure 4.
Schematic of the parallel HP-ORC system with GHE thermal storage capability.

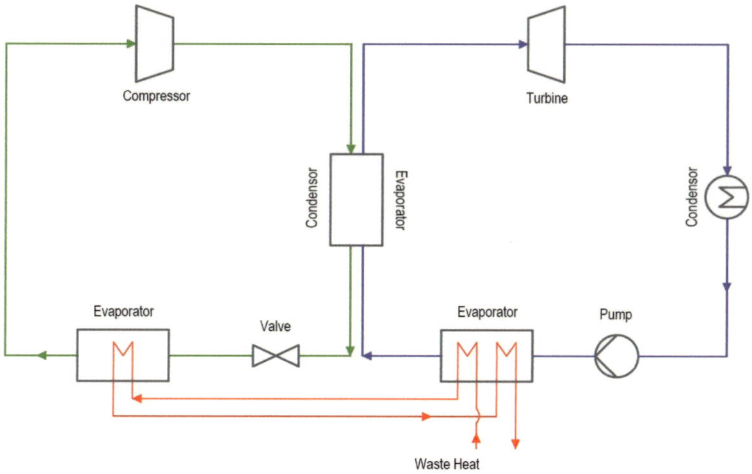

Figure 5.
Schematic of the directly-combined HP-ORC system: (left) HP; (right): ORC.

Research carried out by Mounier [36] investigated ORC-driven HPs also, with the compressor-turbine unit (CTU) being directly coupled the same as the heat exchanger unit shared between the HP and ORC, as depicted in **Figure 6**. A research paper by Collings et al. [37] assessed an identical system with a combined compressor-turbine unit and an air-source HP.

2.3 Reversible HP-ORC systems

A reversible HP-ORC unit is almost identical essentially to a parallel HP-ORC system, where there is versatility within the operations in what is currently performing, except that it integrates the ORC and HP into one unit that can operate in ORC or HP mode conditional on the need. This unit is favourable as it enables for the re-usage of components between these two modes. It additionally makes the use of thermal storage easier, wherein surplus heat can be utilised by the ORC mode to generate electricity with the residual heat stored for use by the HP to utilise for heating as well as hot water intents.

A reversible HP-ORC system has been suggested by Dumont et al. [38, 39], where the reversible unit is coupled with a solar absorber on one side, with the GHE and thermal storage tank being on the other, for ORC mode and HP mode separately. It is interesting to note that this unit moves heat one way, i.e. from the solar-based absorbers toward thermal energy storage. This system additionally gives the ability to the solar powered absorber to directly heat the thermal storage tank in the event that it can give an adequately high temperature; if not, the HP can extract more when necessary.

The system developed by Schimpf and Span [40, 41] was more basic, with the reversible unit being attached to a GHE on one side, and either the solar absorber array or storage tank on the other part as shown in **Figure 7**. In this system, the reversible unit alters path contingent upon the situation, with the ORC mode

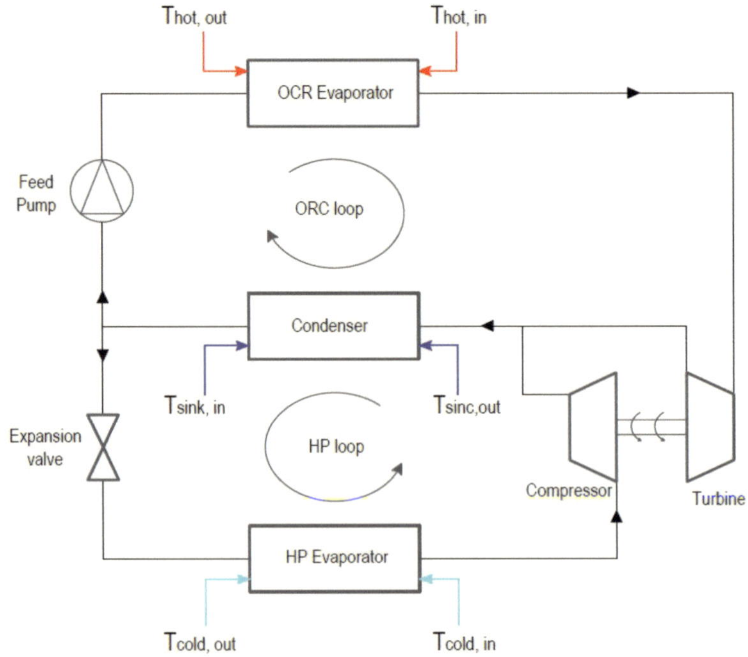

Figure 6.
Schematic of the directly-combined HP-ORC system showing the connected compressor-turbine unit.

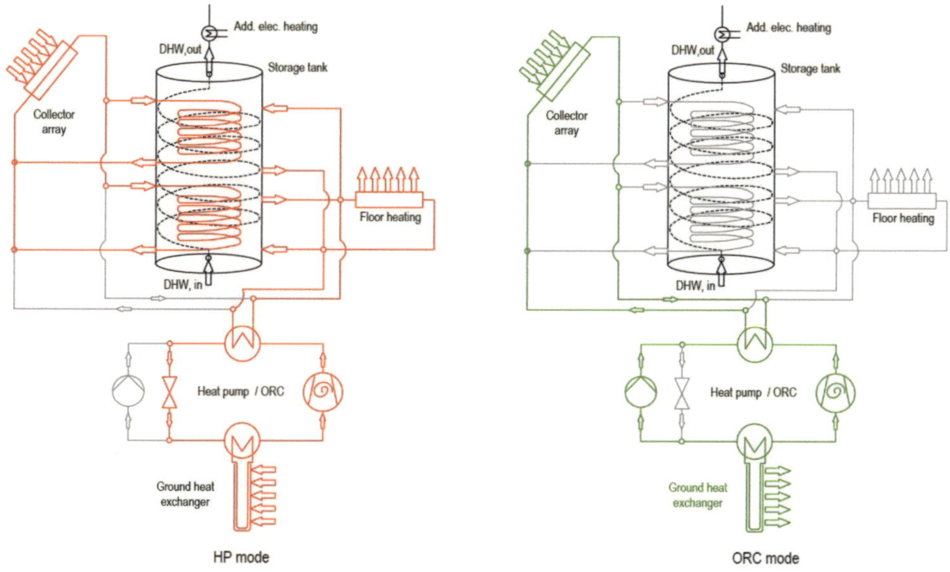

Figure 7.
Schematic of the reversible HP-ORC system: (left) the HP mode; (right) the ORC mode.

transferring heat from the solar panels to the GHE and the HP mode moving heat from the heat exchanger to the storage tank, and in this circumstance, the collector panels directly heat the storage tank too.

3. Performance

The following section on how the combined HP-ORC systems perform will go into details about configurations and results including challenges from the same literature.

3.1 Key performance metrics

As seen before, the essential duty of HP and ORC systems are to supply heating and power, with a considerable lot of these systems moreover performing capacities to supply cooling and domestic hot water. Accordingly, the main performance metrics referenced in these works are aimed at accomplishing these purposes.

There are two central methods to move toward such examination; a subsystem perspective wherein the emphasis is on the effectiveness of the HP-ORC itself, or a whole system point of view where the system's whole capacity is investigated, habitually on a yearly basis. The last is explicitly common when the heat sources and heat loads are transient. The main key performance metrics are listed in **Table 1**. Note that this list is not inclusive, rather only typical of commonly applied metrics.

3.2 Indirectly-combined series

As stated earlier on, these varieties of systems are usually designed for facilitating a greater level of heat recovery from the system, reusing waste heat via processes. In that capacity, numerous systems applicable in this class are from industrial processes and not appropriate for residential utilisation. However, one possibly applicable investigation by Liu et al. [31] recommended a gas engine-powered HP and ORC

Metric name	Typical units	Equations
Subsystem metrics		
Coefficient of performance (heat pump only)	—	For heating: $COP = \frac{Q_H}{W}$ For cooling: $COP = \frac{Q_C}{W}$ $Q_H = heating\ provided$ $Q_C = cooling\ provided$ $W_c = work\ consumed$
Thermal efficiency	%	For heat pump: $n_{th} = \frac{COP}{COP_{rev}}$, $COP_{rev} = \frac{T_H}{T_H - T_C}$ For ORC: $n_{th} = \frac{W_p}{Q_H}$ $W_p = work\ produced$ $T_C = heat\ sink\ temperature$ $T_H = heat\ source\ temperature$
Exergetic efficiency	%	$n_{ex} = \frac{n_{th}}{n_{max}}$, $n_{max} = 1 - \frac{T_C}{T_H}$
Output power	kW	Derivative of work produced by the organic Rankine cycle
Output heat	kW	Derivative of heating provided by subsystem (HP or ORC) to house or another subsystem
Output cooling	kW	Derivative of cooling provided by subsystem (HP or ORC) to house or another subsystem
Whole system metrics		
Net electricity consumption	kWh	Integral of net power generated by system $Net\ power = P_{ORC} - P_{con}$ $P_{ORC} = Power\ generated\ by\ ORC$ $P_{con} = Power\ consumed\ (pumps, etc.)$
Total heating provided	kWh	Integral of heat power output
Total cooling provided	kWh	Integral of cooling power output
Reduction in electricity	%	Difference between electricity consumed by one system compared to a baseline, divided by the baseline's electricity usage
Reduction in emissions	%	Difference between emissions generated by one system compared to a baseline, divided by the baseline's emissions
Total waste heat recovery	%	$Waste\ heat\ recovery = \frac{Q_C}{Q_H}$

Table 1.
Key performance metrics of HP-ORC systems.

system utilising this waste heat recover as shown in **Figure 3**. Experiments of this system confirmed a waste heat recovery of over 55% and have possibility for residential buildings. The total cooling capacity ran between 25 and 48 kW, expanding with higher gas engine speeds. This setup found that as the water delta temperature varied in the range 11.8–24°C, the heat pump COP expanded, however the COP likewise reduced with higher gas engine speeds, running from an estimation of 6.5–10.

3.3 Indirectly-combined parallel

There are a number of potential designs within the parallel type of HP-ORC systems. Most of these arrangements have the HP and ORC units totally apart from each other, which do provide advantages, however, will not be explored here as they can be treated as separate, unintegrated systems. One exclusive parallel design is where the HP and ORC share the GHE, as represented in **Figure 4** [32]. This ORC, utilising R123 as a working fluid and a flowrate of 0.15 kg/s, generated around 2.1–2.2 kW of electricity, rejecting around 19.4–20 kW for heating. The heat pump supplied 24.9–28.7 kW of heating, spending approximately 6.9 kW.

This seems to present further advantages compared to the ground-source HP on its own, through a reduction in power consumption per unit heating area of $2.2 * 10^3$–$3.3 * 10^2$ kWh/m^2, with the ORC unit supplying 55.6% of the total heating capacity, and balancing for 78.5% of the power utilisation of the HP. This more energetically efficient system was capable to better swap a standard GSHP system, decreasing the operation time of the GSHP whilst preserving sustainable ground temperatures that operate in colder climates.

3.4 Directly-combined series

For the directly-combined series systems, one study observed that for a system of this type to be beneficial, some prerequisites are required [33].

a. the evaporation temperature of the ORC is set correctly;

b. the working fluid of the ORC has a slight ratio of latent to sensible heat;

c. poor thermal match between working fluid and waste heat for standalone ORC; and

d. the COP of the HP is adequate.

Applying these settings and the optimisation of heat exchanger temperatures, their model, with an ORC evaporation temperature of 120°C, an ORC condensation temperature to the HP of 45°C and a HP condensation temperature of 130°C, stated an expansion in net power yield and level of waste heat recuperated by 9.37 and 12.04% individually, when utilising n-Hexane as the HP refrigerant and R600a as the ORC working fluid [33]. This system had a power production of 400–800 kW when considering the working liquid, with waste heat recovery between 7000 and 9000 kW.

In a same system utilising solar-based parabolic trough thermal as a heat source for an ORC unit associated with an absorption HP, the greatest power generation attained was 152.1 kW, with the most extreme cooling generation of 465.2 kW. This was accomplished with a working pair of LiBr-H2O, water/steam as the refrigerant, the heat pump absorber and condensers working at 50°C to provide usable heat in a sensible temperature level for space heating or DHW (domestic hot water) outputs, and the HP evaporator working at 10°C to feed the cooling load. The simulation outcomes recommend that the optimal arrangement has an ORC working fluid of toluene, giving heat source temperatures near the extent of 300°C, a greatest exergetic efficiency of 24.66%, pressure ratio of 0.7605 and a heat rejection temperature of 113.7°C [35].

3.5 Reversible HP-ORC

A few investigations have called attention to focal points to exploiting this system, explicitly because it enables for efficient operation in both cold and hot climate conditions using the reversible unit and thermal storage, just as diminishing initial capital expenses. One investigation found that contrasted with the HP solar thermal system employed as the reference, reversing the HP into an ORC unit had the option to diminish the net power request of the system by 2–10% [40]. In this configuration, the ORC mode, with a working fluid of R134a, produced a daily total of between 43.3 and 145.6 kWh, contingent upon the area for solar thermal and the collector type. This production and resulting decrease in net power demand was assisted with the heat pump power between 4.58 and 6.49 kW, 0.2 m^3 of hot temperature water demanded at 45°C and a tank volume of 0.9 m^3.

Another model made established that in winter, the HP spent 17.28 kWh every day with a daily mean COP of 4.1 and throughout the summer, the ORC mode had a 5.5% effectiveness, attaining a peak power of 3.28 kW and producing 23.9 kWh the day it was attained [41]. The main difficulty of this recommended system is choosing the optimal control strategy and decreasing the overall thermal system loss as there are a couple of loss situations that ought to be reduced properly. For example, if the solar loop fluid temperature is adequately high, it may not be viable transmitting it for heating in the mid-year/summer since the house and storage are both at temperature; here the energy will rather be lost. As will be discussed in a later section, trial results founded on these designs are encouraging, showing results fundamentally the same as those simulated.

4. Design optimisation

This section will focus on elements for optimising the overall architecture of the combined energy systems, by considering HP and ORC components design, selection of working fluids, control strategies, and operating temperatures, and managing more variable seasonal temperatures.

4.1 HP and ORC components

From the viewpoint of single system and component optimisation, there are many studies concentrating on HPs or ORCs individually. Because of this, the extent of this section will be abridged where doable to considering studies including components in the framework of integrating HP and ORC into an overall combined heating and power system and their design concerns.

4.1.1 Heat source and sink selections

The differences in temperature between each evaporator and condenser pair is a significant factor in the effectiveness of individual cycle, because it closely influences the notional maximum generation and COP. Unfavourably, because of the residential or commercial uses of these systems, the accessible heat sources are habitually lower temperature, which implies the conceivable temperature differential and resulting effectiveness will be smaller. There are different strategies to enhance or optimise the temperature differential, which will be examined beneath.

For a HP-ORC system combined at the HP condenser/ORC evaporator, a work process to upgrade the working states of the combined system was made [33]. This involves adjusting the combined heat exchanger temperature and computing the other temperatures appropriately, diminishing the combined heat exchanger temperature until it either supplies an acceptable measure of waste heat recovery and expands the net power, or it is anything, but a productive combination. This operating procedure proposes that a diminished coupled heat exchanger temperature will expand the quantity of waste heat recuperated yet may not definitely influence the net power yield, which relies to a great extent upon the optimal or accessible heat source temperature.

For the directly combined compressor-turbine unit design, one investigation established that at a hot source temperature of 120°C, the maximum COP attainable by the comparing HP-ORC framework is 1.66 [36]. For a hot source temperature of 180°C, the maximum COP attainable by this HP-ORC system is above 1.8, with exergetic efficiencies in surplus of half. This analogous analysis concluded that overall, the conventional sorption systems, for example, the single effect absorption HP, function well at low heat source temperatures under 120°C, whereas these rotor-coupled HP-ORC systems function greater at temperatures above 150°C.

While working fluids will be examined in a subsequent section, it should be pointed out that a working fluid that is chosen for use in the system should correspond to the mix of source and sink temperatures for better thermodynamic maximisation.

The configuration of the system will aid prescribe the temperature differentials. Specifically, appropriate choice of a heat source is significant for defining the conceivable temperature from it, and will probably guide the configuration of the whole system.

Combustion heat sources, for example, natural gas and diesel, can normally give greater temperatures yet at a greater emissions generation. These emissions can be mostly decreased by the utilisation of biomass as alternative fuel. Moreover, the dissipated heat from these heat sources can be partly recuperated for additional heat transfer uses in the system as displayed in **Figure 3** [31]. Thus, to the exhaust heat from conventional fuels, the exhaust heat from a SOFC (solid-state fuel cell) can be recovered in a HP-ORC system as the heat source as was investigated in [42] and shown in **Figure 8**, giving a 3–25% increase on exergy efficiency contrasted with the SOFC power cycle, as it stood alone.

One possibility to additionally decrease emissions in comparison with combustion heat sources is via further renewable sources. Aside from electrically heated heat sources, which would have the option to consume power from the grid or nearby sustainable sources, both geothermal and solar thermal alternatives are frequently utilised to give a heat source, taking into consideration a progressively environmental activity, generally speaking. There are a broad assortment of designs

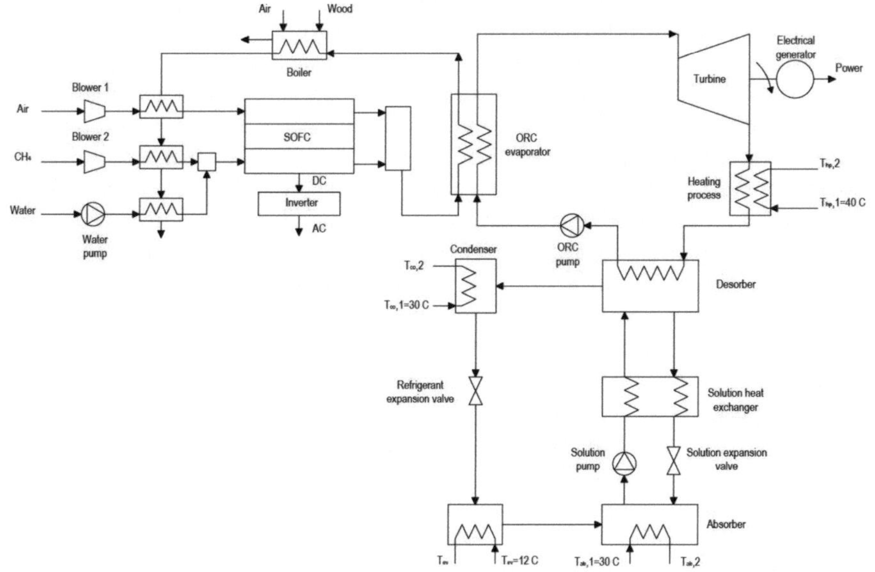

Figure 8.
Schematic flow diagram of the ORC-driven absorption HP system driven by waste heat from a SOFC.

for both of these sources that will not be explored in the extent of this chapter, yet they eventually both go about as heat exchangers for a working fluid experiencing their piping components. Aside from the regional climate conditions, the design of the solar thermal collectors for the most part prescribe the workable heat transfer attainable, with further complex as well as costly designs allowing higher temperatures. Because of their sporadic use, solar thermal heat sources frequently involve an intermediate storage between the ORC-HP circuit and the solar thermal collectors, and may additionally require optional heating capacities if the collectors are not adequate for dependable function.

A number of systems exploit certain type of thermal storage, both for keeping up a required heat source or sink temperature and for saving surplus heat. One basic design is a liquid storage tank, which, if the liquid is water, has the additional capacity of giving residential hot water also. An investigation operating a reversible HP-ORC unit found that for sizing this water storage tank, with estimated temperature extends between 100 and 150°C, the required ORC power yield and release period had a definite correlation to the size of the storage tank, with the temperature differential, examined above, showing an opposite relationship to it [43].

Small-scale geothermal units have, for the most part, lower temperatures by design and as such needs a HP to separate the accessible heat. Even though it is accessible dependably, over the lifespan of the system, the quality of this source will deteriorate when more heat is removed. So to alleviate this, it is conceivable to utilise a geothermal heat exchanger as a type of thermal storage in situations when the heat at a given instant is not essential for the consumer, recovering the geothermal heat source.

This technique was applied in the parallel framework displayed in **Figure 4**. It was discovered that the recharging of the GHE by the ORC had advantages for the complete system when contrasted with just the ground-source HP system, keeping up a greater yearly average COP of 3.8 instead of 3.7 to 3.2 in 20 years, and an 85% decrease in total power utilisation [32].

4.1.2 Expander/compressor units

One exclusive arrangement of this component can be found when the compressor of a HP and expander of the ORC are directly combined across their rotor. This provides the capacity of directly exploiting the mechanical energy from the ORC to power the compression in the HP, in spite of the fact that it represents some mechanical difficulties and involves the two units to be working reliably and dependably for suitable application. The most crucial piece of this system is the turbomachinery, which, in one investigation, when it was enhanced, provided efficiencies in surplus of 60%, a 20 point effectiveness increase compared to their proof of concept, featuring the requirement to limit fluid leak and turbomachinery tip clearances during the fabrication [36]. This work proposed a maximum HP COP amount of 1.8, and for the overall system, had 40 kW heating capacity.

Another research utilising this expander/compressor unit found that a fuel-to-usable heat efficiency of 136–160% was attained, with the HP COP extending between 2.8 and 5.5 [37]. This investigation employed the condensers of the two units to heat up water for household hot water production with the ASHP rising the water temperature to 25–30°C, the ORC condenser expanding it up to 55°C and the exhaust gas from the natural gas combustion for the ORC evaporator expanding the water temperature to 60°C for use in DHW application. The natural gas combustion rises the ORC working fluid evaporator inlet temperature to 200°C, bringing about an ORC efficiency of 20%. Below 5°C of ambient air temperature, 3.9 kW of energy was generated from the ORC to water, compared with 3.8 kW heating from the HP.

Another type of this unit is as a reversible compressor/expander that can work in either way in the event that HP mode or ORC mode is required. One investigation that attempted an assortment of small-scale below <5 kW, expander/compressor units observed that while the biggest isentropic effectiveness was 81% for the scroll expander, compared with 53% for the piston and screw expanders, the mechanical restrictions and working settings are essential for choosing the unit suitable for the application, and recommended an approach to precisely design these units [44].

4.2 Working fluids selection

The selection of working fluid is significant as it directly influences the thermal specificities of the system, for this reason it should be appropriately matched with the required functioning conditions. Working fluid selection is reliant to system arrangement and component sizing. One investigation testing the impact of working fluid choice on system performance concluded that working fluids that have greater decomposition temperatures have greater fuel-to-heat efficiency [45]. Frequently, there are a few working fluids that are comparable in performance of the system. Thus, some compromises must be done. For a rotor-coupled HP-ORC system, one research found that R134a and R152a were the ideal working fluids for this system, with the best selection eventually being an accommodation between the COP and capital expenses [46].

So also, the trade-off can likewise be between technical and environmental performances. Similar to the case with testing of a gas-driven HP-ORC system, where R123 generated the greatest thermal efficiency and energy efficiency of 11.84 and 54.24%, while trials of R245fa generated lower amounts of 11.42 and 52.25%, however showed lower ozone depletion potential and global warming potentials [31]. It ought to be noticed that the combination of an ejector directly into the evaporator subsystem of an ORC can possibly enhance the thermal performance of the whole system via efficient usage of working fluid phase transition [47–50]. In reference to **Figure 9**, a model of this system found that the cycle can generate 10.78% more power, and recover 19.04% more heat from the system [50].

Mixtures of working fluids can enhance the heat transfer capacities in a heat exchanger by giving a more thermodynamically efficient temperature glide, which can possibly intensify the power production and heat recuperation just as at the same time decreasing expenses and ecological effect contrasted with the more costly or higher effect fluid in the mixture.

There are various advantages to working with zeotropic mixtures of dry and wet working fluids, as exhibited by a modelling and simulation study by Zhu et al. [51], which found that inside an ORC combined with an ejector and HP, these blends brought about a higher temperature glide, power effectiveness, cooling efficiency and coefficient of performance. Worth mentioning was R141b/R134a (55:45), R123/R152a (85:15), and R141b/R152a (80:20), which had, individually, the most power yield with a greatest power efficiency of 6%, the most elevated cooling impact with a maximum cooling efficiency of 20.3%, and the maximum COP of 1.18. An evaluation of the relative net power output of working fluid mixtures compared to the best pure working fluids indicate that the potential rise from simulation, ranges from 2.56 to 13.6% relying upon the approach utilised to evaluate this enhancement [52]. It likewise proposed that if big heat exchangers are possible to utilise, the benefits of mixtures will be further obvious.

Furthermore, mixtures of working fluids can be finely adjusted in climate-reliant systems in accordance on the optimal composition for thermodynamic enhancement. A new investigation of an ORC system with composition modification capabilities, has demonstrated this capacity to upgrade the working fluid

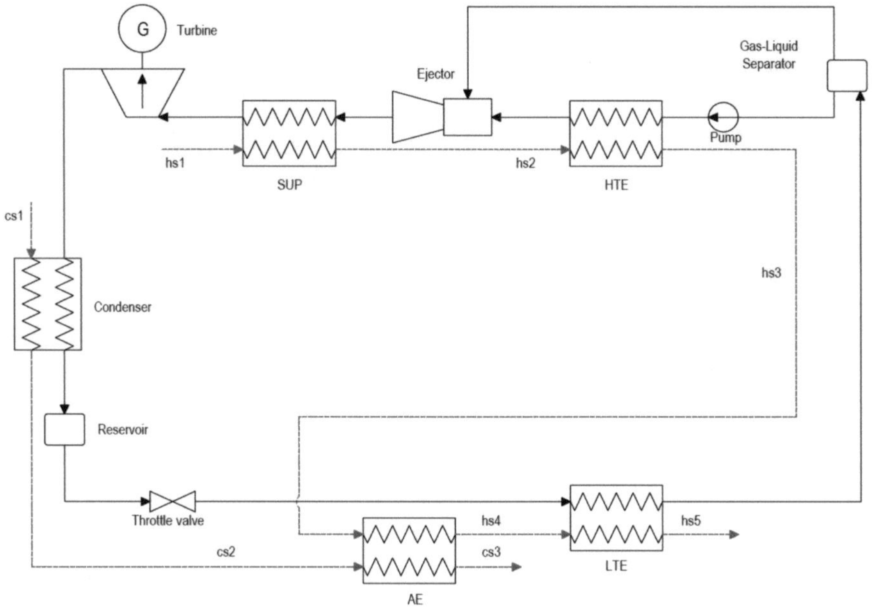

Figure 9.
Schematic diagram of the coupled ejector-ORC. AE: adaptive heat exchanger; HTE: high-temperature heat exchanger; G: generator; LTE: low-temperature heat exchanger; SUP: super heater; cs: cold source; hs: hot source.

composition to best suit surrounding conditions, providing an improvement of yearly mean thermal efficiency by up to 23% over a traditional ORC at a heat source temperature of 100°C, at a rise in capital expense of under 7%, overall proposing smaller reimbursement terms, particularly in areas with big air temperature variances amongst winter and summer periods [53].

4.3 Control strategies

A rise in convolution of a combined HP-ORC system usually requires several control strategies for appropriate arrangements of components and process flow layouts. These control strategies are commonly consisted of thermocouples to check temperature at various positions in the system, choosing the optimal arrangement of components dependent on thermal accessibility and requirement. A variant of this can be found in the system delineated by **Figure 4** where a duty cycle controls when it is supposed the heating season, when the red stream channels are operating, and non-heating season, when the green streams are effective [32]. During the heating season, a basic differential temperature controller controls if the HP is on or off.

Additionally, the reversible system displayed in **Figure** 7 exploit both solar thermal and ground-source HP to expand the temperature in a thermal storage tank. When the tank attains the required temperature, any surplus solar energy is transferred to the now-reversed HP in ORC mode, where it is consumed to generate power and recharge the GHE [40, 41].

While most of systems considered apply the temperature monitoring to recommend system arrangement and system conditions, there are a couple of works that have tried further innovative control. One investigation examined the impacts strategies, for example, better space heating control when household hot water is needed, and the monitoring of occupancy to adjust set point temperatures, and at last established that they are typically beneficial in enhancing thermal and electrical efficiency and decreasing friction on the system parts [54]. Nonetheless, it was noticed that their exploitation and viability do differ dependent on the building properties and occupant schedule, and ought to be applied with attention.

4.4 Variable weather effect

Generally, if there were slight difference in the climate, both hot and cold temperature areas would not have to modify their combined HP-ORC system arrangement to satisfy seasonal conditions. Unfavourably, in areas that have much changeable climate (akin to a lot of Canada with cold winters and hot summers), seasonal prerequisites would prescribe an adjustment in the HP-ORC system arrangement. Therefore, both reversible HP-ORC systems and parallel units are best alternatives, as the adaptability in these systems take into account such modifications in arrangement, permitting indoor temperatures to be kept up during the time using thermal storage, and efficient component control.

A significant number of these suitable systems have been surveyed in earlier sections, and will just be referenced here. One investigation demonstrated an ORC with working fluid composition fine-tuning, permitting the system to be thermodynamically maximised dependent on the climate conditions [53]. A parallel system was simulated in an area of China with comparable temperature variances to Ottawa, Ontario, and incorporated a GHE thermal storage with recharging options within its design and control strategy [32]. At last, while not expressly examined in reference to big temperature differences, the investigations with reversible HP-ORC systems address the theme indirectly by their but milder areas of Belgium, Denmark and Germany [40, 41, 43, 44].

5. Experimental studies

Most of the studies done on these combined systems has been carried out by means of modelling and simulation studies, for example, the ones considered previously. Generally, this is reasonable, since several of the separable components in the system have validated models to achieve a specific understanding of the system dynamics and viability. There have been a couple of proofs of concept for a combined HP-ORC framework. While it will not be provided directly, individual ORC systems have been investigated, validating past thermodynamic models compared to experimental ORC designs, for example, ORCmKit [55]. As ORC units are generally more recent products, the assembling and configuration needed will probably bring about some intrinsic changeability in results compared with the models.

Even though the performance outcomes were before addressed in detail, the work with gas engine-driven HP and ORC heat recovery unit introduced results on a test rig of the system, developing their thermodynamic model of the ORC. For their model of the system, it was discovered that the greatest uncertainty for cooling capacity was 1.23%, gas engine energy consumption was 0.57%, waste heat was 2.11%, COP was 3.42% and primary energy ratio was 3.56%, all proposing a reasonably high conformity amongst their created models and the experimental design [31].

For the heat exchanger-combined HP-ORC unit, when contrasting their created models to proof of concept experimental data, it was revealed that their proof of concept has a 30% lesser COP and 43% greater expenses than simulated, recommending that these losses are because of not well optimised compressor-turbine units and heat exchangers and can be improved essentially with appropriate refinement [36].

While there have been works trying separate HPs or ORCs, an investigation regarding the reversible HP-ORC system analysed the experimental results from the system, finding different items for real application. This study established that a cycle efficiency of 4.2% is accomplished in ORC mode, from condensation and evaporation temperatures of 25 and 88°C individually, and a COP of 3.1 being acquired in HP mode from condensation and evaporation temperatures of 61 and 21°C, respectively, demonstrating the viability of the concept [38]. One cause of efficiency loss happened at the expander/compressor, as it was not at first geometrically intended for reversible application.

5.1 Experimental works

For model development, validation, and component optimisation intents, it is usual to disconnect or apart a subsystem to assist assessing its viability and use in the more extensive system preceding any bigger scale testing or demonstration. For these separate subsystems, a method known as the reconciliation method can be applied, which means to characterise the most likely physical condition of a system and modify every estimation as much as possible through information on its precision, duplications, restrictions, and solving mass and energy balances [56]. Through this investigation, exploiting this technique when implemented to a reversible HP-ORC unit allowed further effective data collection and validation, decreasing the error for example in the situation of the pinch-point calculation of an evaporator where its normalised root mean square deviation was diminished from 14.3 to 4.1%.

5.2 Deployment

There has not been any associated cases of combined HP-ORC systems beyond experimental facilities for building applications. From the past modelling and experimental studies, in any case, there are an assortment of concerns that ought to be examined once demonstrating these systems. One of these matters is the relative novelty of ORC units. These units, particularly at a small and micro scale, have not generally been adequately optimised in real-word use, and there is lesser data about it contrasted with well-known systems. This will in general reduce the possible advantages and rise the whole prices owing partly to maintenance necessities.

Another huge concern is the territorial changeability in climate and temperature. All together for the system to be monetarily attainable, the system must be designed explicitly to suit the consumers' requirements and application needs, which will be affected by the climate existent in the area. There must likewise be an assurance on what capacities are wanted such as cooling, heating, domestic hot water, power, which will likewise determine the arrangement of the system.

6. Economic analysis

For economic analysis, the basic equations applied are from standard engineering economics. In HP-ORC projects, the determinants of costs comprise of the underlying initial capital for buying the equipment, the net power consumption, and the related overhaul and maintenance necessities. In light of these qualities, the yearly prices and savings can be viably decided, taking into consideration an income investigation, evaluation of return and also financial viability. These outcomes enable for concrete examination between potential systems.

Numerous numbers of the studies realised concerning combined HP-ORC systems cover some financial analysis. Because of the intermittence of heating and power options, it is challenging to assess undeviating comparison amongst them accordingly. In contrast with a solar thermal system combined to a ground-source HP, one investigation assessed that the adaptation of this system to a reversible HP-ORC system for a residential system would just cost roughly $600, and following 20 years of utilisation, presents benefits of $230 in Ankara and $110 in Denver [40, 41]. Moreover, it recommends that the primary factors for the viability of the system is the area, especially continental climate conditions, in addition to components and pump expenses, including the working fluid choosing, which the pump is suitable with.

In view of the reversible HP-ORC system designed by Dumont [44], a comparison of this system with the further developed HP and PV (photovoltaic) system, indicates that as a whole, the reversible system is fewer beneficial, albeit an expansion in heat demand for heating or domestic hot water, can possibly enhance the effectiveness of the reversible system over that of the HP and PV system. This is additionally established from temperature areas with similarly lower temperature deviations. It prescribes further investigation is performed to ascertain where the reversible system is financially cost-effective against the HP and PV system.

A few models created stated reimbursement periods for the PV-ORC-chiller arrangement to be 9 years and 6.5 years in Amsterdam and Athens, individually, and for the combined solar thermal-ORC-chiller to have return periods of 16.5 years and 49.5 years in similar areas [34]. Likewise, the greatest exergy efficiency of the described models, for the PV case is 2 and 6%, while the solar thermal case is 18 and 37%. The sensitivity analysis performed with this work disclosed that if the energy cost increments by 10%, this could decrease the return period by 12.9–15.0%, besides it is the best effective factor on the financial outputs.

As shown, a portion of the bigger causes of changeability amongst implementation, are the power costs and areas climate, with the whole system design at last being advised by these factors. Improvement of this system will have some effect on the financial viability, albeit such changes ought to be assessed cautiously to guarantee some profit is achieved.

7. Conclusions

In conclusion, the combination of HP and ORC units have capacity for more energy efficiencies in various configurations. This survey provided some innovative concepts and designs to aid in prospect modelling and experimental studies alike, pinpointing various issues of more research. The main matters are:

- For improving the heat transfer, it is suggested to emphasise on maximising the temperature differential between the heat source and heat sink, similarly as suitably adjusting a working fluid for the specified ranges. Accounts of the heat sources and comprising thermal storage will be truly efficient at enhancing performance.

- Designs should be selected reliant on what is required from the system. Changing climate conditions across countries such as Canada, would need a variety of heating, cooling, and domestic hot water demands, in addition to being a concern itself for selection of heat sources, such as from air-source and ground-source HPs to solar thermal collectors. Some designs and arrangements are just applicable for particular ambient air temperatures or weather conditions, despite an improved optimal result is to have better control or resilience in the system if there is difference in seasons.

- Other optimisation approaches, for instance, advanced control strategies and individual component optimisation, will have certain influence on upgrading the system, regardless of the way this will not be as substantial as the alternatives above and should be counted economically as it might not be feasible to implement for every circumstance.

- On defining the weather conditions, the area is fundamental at choosing energy costs. Both of these two elements are maybe the greatest influence on assessing economic feasibility and hence, cannot be neglected.

For future research, it is recommended to:

- Compare a variety of advanced designs to a standard to understand which systems are optimal for a range of cold climate (such as Canadian) areas.

- Explore the combination of these systems in newly designs and configurations to realise alternative solutions for utilisation options, for instance, for isolated locations.

- Evaluate relevant combined HP-ORC systems and their readiness for bench testing, demonstration, deployment, in addition to developing models founded on them to well design the systems.

- Perform a sensitivity analysis to understand main drivers in the design and optimisation of systems of interest.

Acknowledgements

Funding for this work was provided by Natural Resources Canada through the Program of Energy Research and Development.

Conflict of interest

The authors declare no conflict of interest.

Nomenclature

AE	adaptive heat exchanger
ASHP	air-source heat pump
COP	coefficient of performance
COP_{rev}	coefficient of performance of reversible HP
CTU	compressor-turbine unit
G	generator
GHE	ground heat exchanger
GHGs	greenhouse gas emissions
GSHP	ground-source heat pump
HP	heat pump

HTE	high-temperature heat exchanger
hs	hot source
LTE	low-temperature heat exchanger
ORC	organic Rankine cycle
PV	photovoltaic
Q_C	cooling provided
Q_H	heating provided
R123	2,2-dichloro-1,1,1-trifluoroethane ($C_2HCl_2F_3$)
R134	1,1,1,2-tetrafluoroethane (CF_3CH_2F)
R141b	1,1-dichloro-1-fluoroethane ($C_2H_3Cl_2F$)
R245fa	1,1,1,3,3-pentafluoropropane is a hydrofluorocarbon ($C_3H_3F_5$)
R152a	1,1-difluoroethane ($C_2H_4F_2$)
R600	n-Butane (C_4H_{10})
SOFC	solid-state fuel cell
SUP	super heater
T_C	heat sink temperature
T_H	heat source temperature
WSHP	water-source heat pump
W_C	work consumed
W_p	work produced

Greek letters

n_{ex}	exergetic efficiency
n_{th}	thermal efficiency

Author details

Wahiba Yaïci[1]*, Evgueniy Entchev[1], Pouyan Talebizadeh Sardari[2] and Michela Longo[3]

1 CanmetENERGY Research Centre, Natural Resources Canada, Ottawa, Canada

2 Faculty of Engineering, University of Nottingham, Nottingham, United Kingdom

3 Department of Energy, Politecnico di Milano, Milan, Italy

*Address all correspondence to: wahiba.yaici@canada.ca

References

[1] Office of Energy Efficiency. Residential Secondary Energy Use (Final Demand) by Energy Source and End Use, National Energy Use Database [Online]. Canada: Natural Resources Canada; 2016. Available from: http://oee.nrcan.gc.ca/corporate/statistics/neud/dpa/showTable.cfm?type=HB&sector=res&juris=00&rn=1&page=0 [Accessed: 04 May 2020]

[2] Tchanche BF, Lambrinos G, Frangoudakis A, Papadakis G. Low-grade heat conversion into power using organic Rankine cycles—A review of various applications. Renewable and Sustainable Energy Reviews. 2011;**15**:3963-3979

[3] Quoilin S, Van Den Broek M, Declaye S, Dewallef P, Lemort V. Techno-economic survey of organic Rankine cycle (ORC) systems. Renewable and Sustainable Energy Reviews. 2013;**22**:168-186

[4] Bao J, Zhao L. A review of working fluid and expander selections for organic Rankine cycle. Renewable and Sustainable Energy Reviews. 2013;**24**:325-342

[5] Ziviani D, Beyene A, Venturini M. Advances and challenges in ORC systems modeling for low grade thermal energy recovery. Applied Energy. 2014;**121**:79-95

[6] Patrick L, Athanasios P, Panos S. Systematic methods for working fluid selection and the design, integration and control of organic Rankine cycles—A review. Energies. 2015;**8**:4755-4801

[7] Lecompte S, Huisseune H, van den Broek M, De Paepe M. Review of organic Rankine cycle (ORC) architectures for waste heat recovery. Renewable and Sustainable Energy Reviews. 2015;**47**:448-461

[8] Rahbar K, Mahmoud S, Al-Dadah RK, Moazami N, Mirhadizadeh SA. Review of organic Rankine cycle for small-scale applications. Energy Conversion and Management. 2017;**134**:135-155

[9] Tocci L, Pal T, Pesmazoglou I, Franchetti B. A small scale organic Rankine cycle (ORC): A techno-economic review. Energies. 2017;**10**:413

[10] Mahmoudi A, Fazli M, Morad MR. A recent review of waste heat recovery by organic Rankine cycle. Applied Thermal Engineering. 2018;**143**:660-675

[11] Park BS, Usman M, Imran M, Pesyridis A. Review of organic Rankine cycle experimental data trends. Energy Conversation and Management. 2018;**173**:679-691

[12] Pereira JS, Ribeiro JB, Mendes R, Vaz GC, André JC. ORC based micro-cogeneration systems for residential application—A state of the art review and current challenges. Renewable and Sustainable Energy Reviews. 2018;**92**:728-743

[13] Xu W, Zhao L, Mao SS, Deng S. Towards novel low temperature thermodynamic cycle: A critical review originated from organic Rankine cycle. Energy. 2019;**167**:484-497

[14] Chua KJ, Chou SK, Yang WM. Advances in heat pump systems: A review. Applied Energy. 2010;**87**:3611-3624

[15] Haller MY, Bertram E, Dott R, Afjei T, Ochs F, Hadorn JC. Review of component models for the simulation of combined solar and heat pump heating systems. Energy Procedia. 2012;**30**:611-622

[16] Omojaro P, Breitkopf C. Direct expansion solar assisted heat pumps: A review of applications and recent

research. Renewable and Sustainable Energy Reviews. 2013;**22**:33-45

[17] Sarbu I, Sebarchievici C. General review of ground-source heat pump systems for heating and cooling of buildings. Energy and Buildings. 2014;**70**:441-454

[18] Kamel RS, Fung AS, Dash PRH. Solar systems and their integration with heat pumps: A review. Energy and Buildings. 2015;**87**:395-412

[19] Buker MS, Riffat SB. Solar assisted heat pump systems for low temperature water heating applications: A systematic review. Renewable and Sustainability Energy Reviews. 2016;**55**:399-413

[20] Fischer D, Madani H. On heat pumps in smart grids: A review. Renewable and Sustainable Energy Reviews. 2017;**70**:342-357

[21] Zhang L, Jiang Y, Dong J, Yao Y. Advances in vapor compression air source heat pump system in cold regions: A review. Renewable and Sustainable Energy Reviews. 2018;**81**:353-365

[22] Mohanraj M, Belyayev Y, Jayaraj S, Kaltayev A. Research and developments on solar assisted compression heat pump systems—A comprehensive review (Part-B: Applications). Renewable and Sustainable Energy Reviews. 2018;**83**:124-155

[23] Poppi S, Sommerfeldt N, Bales C, Madani H, Lundqvist P. Techno-economic review of solar heat pump systems for residential heating applications. Renewable and Sustainable Energy Reviews. 2018;**81**:22-32

[24] Dias JMS, Costa VAF. Adsorption heat pumps for heating applications: A review of current state, literature gaps and development challenges. Renewable and Sustainable Energy Reviews. 2018;**98**:317-327

[25] Lorenzo C, Narvarte L. Performance indicators of photovoltaic heat-pumps. Heliyon. 2019;**5**:e02691

[26] Goyal A, Staedter MA, Garimella S. A review of control methodologies for vapor compression and absorption heat pumps. International Journal of Refrigeration. 2019;**97**:1-20

[27] Nouri G, Noorollahi Y, Yousefi H. Solar assisted ground source heat pump systems—A review. Applied Thermal Engineering. 2019;**163**:114351

[28] Wang X, Xia L, Bales C, Zhang X, Copertaro B, Pan S, et al. A systematic review of recent air source heat pump (ASHP) systems assisted by solar thermal, photovoltaic and photovoltaic/thermal sources. Renewable Energy. 2020;**146**:2472-2487

[29] Pinheiro M, Salústio S, Rocha J, Valente AA, Silva CM. Adsorption heat pumps for heating applications. Renewable and Sustainable Energy Reviews. 2020;**119**:109528

[30] Yaïci W, Entchev E, Talebizadeh Sardari P, Longo M. Heat pump-organic Rankine cycle hybrid systems for co/tri-generation applications: A state-of-the-art overview. In: Proceedings of the ASME 14th International Conference on Energy Sustainability, 16, 17 June 2020. New York, USA: American Society of Mechanical Engineers (ASME); 2020

[31] Liu H, Zhou Q, Zhao H, Wang P. Experiments and thermal modeling on hybrid energy supply system of gas engine heat pumps and organic Rankine cycle. Energy and Buildings. 2015;**8**:7226-7232

[32] Li W, Lin X, Cao C, Gong Z, Gao Y. Organic Rankine cycle-assisted ground source heat pump combisystem for space heating in cold regions. Energy Conversion and Management. 2018;**165**:195-405

[33] Yu H, Gundersen T, Feng X. Process integration of organic Rankine cycle (ORC) and heat pump for low temperature waste heat recovery. Energy. 2018;**160**:330-340

[34] Roumpedakis T. Techno-economic investigations of a solar driven ORC-sorption system for combined cooling, heating and power [MSc thesis]. Delft, the Netherlands: TUDelft; 2018

[35] Bellos E, Tzivanidis C. Optimization of a solar-driven trigeneration system with nanofluid-based parabolic trough collectors. Energies. 2017;**10**:848

[36] Mounier V. Potential and challenges of ORC driven heat pumps based on gas bearing supported turbomachinery [PhD thesis]. Lausanne, Switzerland: École Polytechnique Féderale de Lausanne; 2018

[37] Collings P, Al-Tameemi M, Yu Z. A combined organic Rankine cycle-heat pump system for domestic hot water application. In: Proceedings of the 12th International Conference on Heat Transfer, Fluid Mechanics and Thermodynamics. Malaga, Spain, 11-13 July 2016. Glasgow, UK: Enlighten Editors; 2016

[38] Dumont O, Quoilin S, Lemort V. Experimental investigation of a reversible heat pump/organic Rankine cycle unit designed to be coupled with a passive house (net zero energy building). International Journal of Refrigeration. 2015;**54**:190-403

[39] Dumont O, Carmo C, Randaxhe F, Quoilin S, Lemort V. Simulation of a passive house coupled with a heat pump/organic Rankine cycle reversible unit. In: Proceedings of the 9th International Conference on System Simulation in Buildings. Liège, Belgium: University of Liège; 2014

[40] Schimpf S, Span R. Techno-economic evaluation of a solar assisted combined heat pump—Organic Rankine cycle system. Energy Conversion and Management. 2015;**94**:430-437

[41] Schimpf S, Span R. Simulation of a novel solar assisted combined heat pump—Organic Rankine cycle system. In: Proceedings of the 6th International Conference on Applied Energy, 30 May–2 June 2014. Taipei, Taiwan: Elsevier; 2014

[42] Al-Sulaiman FA, Dincer I, Hamdullahpur F. Exergy analysis of an integrated solid oxide fuel cell and organic Rankine cycle for cooling, heating and power production. Journal of Power Sources. 2010;**195**:2346-2354

[43] Staub S, Bazan P, Braimakis K, Muller D, Regensburger C, Scharrer D, et al. Reversible heat pump–organic Rankine cycle systems for the storage of renewable electricity. Energies. 2018;**11**:1352

[44] Dumont O. Investigation of a heat pump reversible into an organic Rankine cycle and its application in the building sector [PhD thesis]. Liege, Belgium: University of Liège; 2017

[45] Liang Y, Yu Z. Working fluid selection for a combined system based on coupling of organic Rankine cycle and air source heat pump cycle. In: Proceedings of the 10th International Conference on Applied Energy, 22-25 August 2018. Hong Kong, China: Elsevier; 2018

[46] Mounier V, Schiffmann J. ORC Driven Heat Pump Running on Gas Bearings for Domestic Applications: Proof of Concept and Thermo-Economic Improvement Potential. Rotterdam: International Energy Agency; 2017

[47] Wang J, Dai Y, Gao L. Parametric analysis and optimization for a combined power and refrigeration cycle. Applied Energy. 2008;**85**:1071-1085

[48] Wang J, Dai Y, Sun Z. A theoretical study on a novel combined power and ejector refrigeration cycle. International Journal of Refrigeration. 2009;**32**:1186-1194

[49] Wang J, Dai Y, Zhang T, Ma S. Parametric analysis for a new combined power and ejector/absorption refrigeration cycle. Energy. 2009;**34**:1587-1593

[50] Zhang C, Lin J, Tan Y. A theoretical study on a novel combined organic Rankine cycle and ejector heat pump. Energy. 2019;**176**:81-90

[51] Zhu Y, Li W, Sun G. Thermodynamic analysis of evaporation temperature glide of zeotropic mixtures on the ORC-CCHP system integrated with ejector and heat pump. In: 10th International Conference on Applied Energy. Hong Kong, 2018. Energy Procedia. 2019;**158**:1632-1639

[52] Andreasen JG, Kaern MR, Haglind F. Assessment of methods for performance comparison of pure and zeotropic working fluids for organic Rankine cycle power systems. Energies. 2019;**12**:1783

[53] Collings P, Yu Z, Wang E. A dynamic organic Rankine cycle using a zeotropic mixture as the working fluid with composition tuning to match changing ambient conditions. Applied Energy. 2016;**171**:581-591

[54] Carmo C, Nielsen MP, Elmegaard B, Dumont O. Performance evaluation of a HP/ORC (heat pump/organic Rankine cycle) system with optimal control of sensible thermal storage. In: Proceedings of the 4th International High Performance Buildings Conference, 14 July 2016. West Lafayette, Indiana, USA: Purdue; 2016

[55] Dickes R, Ziviani D, De Paepe M, van den Broek M, Quoilin S, Lemort V. ORCmKit: An open-source library for organic Rankine cycle modelling and analysis. In: The Proceedings of ECOS 2016: The 29th International Conference on Efficiency, Cost, Optimization, Simulation and Environmental Impact of Energy Systems. Portorož, Slovenia: University of Ljubljana, Faculty of Mechanical Engineering; 2016

[56] Dumont O, Quoilin S, Lemort V. Importance of the reconciliation method to handle experimental data in refrigeration and power cycle: Application to a reversible heat pump/organic Rankine cycle unit integrated in a positive energy building. International Journal of Energy and Environmental Engineering. 2016;**7**:137-143